Konrad Schwanitz

The TiO2/Dye/Electrolyte-Interface in the Dye Sensitized Solar Cell

Konrad Schwanitz

The TiO2/Dye/Electrolyte-Interface in the Dye Sensitized Solar Cell

A Synchrotron Induced Photoelectron Spectroscopy Study

Südwestdeutscher Verlag für Hochschulschriften

Impressum/Imprint (nur für Deutschland/ only for Germany)
Bibliografische Information der Deutschen Nationalbibliothek: Die Deutsche Nationalbibliothek verzeichnet diese Publikation in der Deutschen Nationalbibliografie; detaillierte bibliografische Daten sind im Internet über http://dnb.d-nb.de abrufbar.
Alle in diesem Buch genannten Marken und Produktnamen unterliegen warenzeichen-, marken- oder patentrechtlichem Schutz bzw. sind Warenzeichen oder eingetragene Warenzeichen der jeweiligen Inhaber. Die Wiedergabe von Marken, Produktnamen, Gebrauchsnamen, Handelsnamen, Warenbezeichnungen u.s.w. in diesem Werk berechtigt auch ohne besondere Kennzeichnung nicht zu der Annahme, dass solche Namen im Sinne der Warenzeichen- und Markenschutzgesetzgebung als frei zu betrachten wären und daher von jedermann benutzt werden dürften.

Verlag: Südwestdeutscher Verlag für Hochschulschriften Aktiengesellschaft & Co. KG
Dudweiler Landstr. 99, 66123 Saarbrücken, Deutschland
Telefon +49 681 37 20 271-1, Telefax +49 681 37 20 271-0, Email: info@svh-verlag.de
Zugl.: Darmstadt, TU, Diss., 2008

Herstellung in Deutschland:
Schaltungsdienst Lange o.H.G., Berlin
Books on Demand GmbH, Norderstedt
Reha GmbH, Saarbrücken
Amazon Distribution GmbH, Leipzig
ISBN: 978-3-8381-0742-4

Imprint (only for USA, GB)
Bibliographic information published by the Deutsche Nationalbibliothek: The Deutsche Nationalbibliothek lists this publication in the Deutsche Nationalbibliografie; detailed bibliographic data are available in the Internet at http://dnb.d-nb.de.
Any brand names and product names mentioned in this book are subject to trademark, brand or patent protection and are trademarks or registered trademarks of their respective holders. The use of brand names, product names, common names, trade names, product descriptions etc. even without a particular marking in this works is in no way to be construed to mean that such names may be regarded as unrestricted in respect of trademark and brand protection legislation and could thus be used by anyone.

Publisher:
Südwestdeutscher Verlag für Hochschulschriften Aktiengesellschaft & Co. KG
Dudweiler Landstr. 99, 66123 Saarbrücken, Germany
Phone +49 681 37 20 271-1, Fax +49 681 37 20 271-0, Email: info@svh-verlag.de

Copyright © 2009 by the author and Südwestdeutscher Verlag für Hochschulschriften Aktiengesellschaft & Co. KG and licensors
All rights reserved. Saarbrücken 2009

Printed in the U.S.A.
Printed in the U.K. by (see last page)
ISBN: 978-3-8381-0742-4

Contents

Abstract   5

Übersicht   7

Introduction to the Dye Sensitized Solar Cell   9

I   Fundamentals   13

1 The TiO_2/dye/electrolyte interface   15
    1.1   Alignment of electronic energy levels 15
    1.2   The electrochemical double layer 17
    1.3   Charge transfer processes 19
          1.3.1   Electron injection 19
          1.3.2   Dye rereduction 20
          1.3.3   Electron transport 21
          1.3.4   Charge recombination processes 26
    1.4   Ongoing development of the DSSC 27

2 The investigated components of the Dye Sensitized Solar Cell   29
    2.1   The TiO_2 substrate 29
          2.1.1   Crystal phases and surfaces 29
          2.1.2   Shape of nanocrystalline crystallites 33
          2.1.3   Electronic structure 34
          2.1.4   Defects 36
    2.2   The N3 Ru dye 39
          2.2.1   Molecular and electronic structure 39
          2.2.2   Adsorption geometry on anatase 41
          2.2.3   Anchoring modes onto anatase 43
    2.3   The electrolyte 45
          2.3.1   Solvation of the electrolyte ions 45
          2.3.2   Electronic structure 47
          2.3.3   The I^-/I_3^- redox couple 48
          2.3.4   The cation 49
          2.3.5   The solvent 50

II Experimental — 53

3 Experimental characterization methods — 55
- 3.1 Photoelectron Spectroscopy (PES) — 55
- 3.2 Raman Spectroscopy — 61
- 3.3 Grazing Incidence X-Ray Diffraction (GIXRD) — 62
- 3.4 Atomic Force Microscopy (AFM) — 62
- 3.5 Scanning Electron Microscopy (SEM) — 64

4 Substrate and interface preparation — 67
- 4.1 TiO_2 substrate preparation — 67
  - 4.1.1 Sol-Gel preparation — 67
  - 4.1.2 Metal Organic Chemical Vapor Deposition (MOCVD) — 68
- 4.2 Interface preparation — 68
  - 4.2.1 Solid-Liquid Analysis System — 69
  - 4.2.2 Dye and electrolyte adsorption — 70
  - 4.2.3 Solvent (co)adsorption — 70

III Results and Discussion — 73

5 Scope of this work — 75

6 The TiO_2 substrate — 77
- 6.1 Crystalline phases — 77
  - 6.1.1 GIXRD — 77
  - 6.1.2 Raman — 79
- 6.2 Morphology — 80
  - 6.2.1 nc-TiO_2 — 80
  - 6.2.2 MOCVD-TiO_2 — 81
  - 6.2.3 Summary — 83
- 6.3 Electronic structure and defects — 83
  - 6.3.1 Survey spectrum of core level emissions — 83
  - 6.3.2 Surface contributions in core level emission spectra — 84
  - 6.3.3 Valence band and band gap — 89
  - 6.3.4 ResPES of bandgap states — 90
  - 6.3.5 Discussion of the results — 95

7 TiO_2/solvent interface — 97
- 7.1 Adsorption of acetonitrile / benzene on TiO_2 — 97
  - 7.1.1 Coverage by the adsorbed solvent — 97
  - 7.1.2 Interaction between solvent and oxygen defects — 100
  - 7.1.3 Work function changes — 104
  - 7.1.4 Solvent-oxygen interaction — 106
- 7.2 Water adsorption — 108
- 7.3 Discussion of the results — 111

8 TiO$_2$/N3 Ru dye interface — 113
8.1 Components of the dye — 113
8.1.1 Carbon and Ruthenium — 113
8.1.2 Nitrogen — 116
8.1.3 Oxygen — 118
8.1.4 Sulphur — 119
8.2 Interaction with the substrate — 121
8.2.1 HOMO and oxygen vacancies — 121
8.2.2 Titanium — 123
8.2.3 Work function — 124
8.3 Discussion of the results — 125

9 TiO$_2$/N3 Ru dye/solvent interface — 127
9.1 Interaction of the coadsorbed solvent with the substrate — 127
9.2 Dye–solvent interaction — 129
9.3 Work function changes — 137
9.4 Discussion of the results — 138

10 TiO$_2$ / iodide / solvent interface — 143
10.1 The two electrolyte components: LiI and molten salt — 143
10.2 Coadsorption of iodide salt and acetonitrile — 146
10.2.1 LiI + acetonitrile — 146
10.2.2 molten salt + acetonitrile — 149
10.3 Discussion of the results — 151

11 TiO$_2$ / N3 Ru dye / iodide / solvent interface — 153
11.1 Coadsorption of N3 dye, molten salt and acetonitrile — 153
11.2 Discussion of the results — 156

12 Summary & Outlook — 159
12.1 TiO$_2$ substrate — 159
12.2 Solvent adsorption — 159
12.3 Dye adsorption — 160
12.4 Dye/solvent coadsorption — 161
12.5 Electrolyte salt/solvent coadsorption — 161
12.6 Dye/electrolyte salt/solvent coadsorption — 162
12.7 Outlook — 163

Bibliography — I

Danksagung — XVII

Abstract

In this work it has been attempted for the first time to investigate the complete working electrode (half cell interface: $TiO_2/RuL'_2(NCS)_2$/electrolyte) of the Dye Sensitized Solar Cell (DSSC) by means of Photoelectron Spectroscopy. The scope of this work is to investigate the interfacial interactions between the adsorbates and the TiO_2 substrate, rather than bulk processes within the substrate or electrolyte.

Since the coverage of the dye onto the TiO_2 substrate amounts only up to one monolayer, a highly surface sensitive method, as Synchrotron Induced Photoelectron Spectroscopy (SXPS), has been applied to investigate the topological, chemical and especially the electronic interactions within the working electrode interface. The photoelectron spectra have been recorded predominantly at the U49/PGM-2 and the TGM7 beamlines at BESSY, and some have been measured at the Darmstadt Integrated System (DAISY). For preparation and analysis of the solid/liquid interfaces a specifically designed experimental workstation (SoLiAS) has been employed, run by the Surface Science Group of Materials Science at the University of Technology in Darmstadt. All investigated interfaces have been prepared in the Ultrahigh Vacuum (UHV) and transferred to the analysis chamber without contamination by ambient air. In order to be able to adsorb the solvent in the UHV, a liquid nitrogen cooled manipulator has been used at SoLiAS.

The adsorption of acetonitrile and benzene onto the untreated TiO_2 substrate reveals the interactions of the solvent molecules with the substrate to a different extent. Whereas acetonitrile quenches Ti^{3+} sites of reduced coordination numbers and oxygen vacancies (V_O states) to a considerable amount, the interaction of benzene is rather weak. In contrast to acetonitrile adsorption, where a fraction of molecules is obviously chemisorbed to the surface, benzene molecules are mostly physisorbed.

Of special interest for the function of the DSSC device is the directional transfer of holes and electrons due to the distinct geometry of the dye molecule, often referred as *vectorial charge transfer* [1]. This implies, that the dye ideally works as an electronic membrane, transferring holes and electrons into opposite directions. Therefore the orientation of the dye and its coupling to the TiO_2 substrate is the crucial point to allow an effective charge separation. By coadsorption of polar acetonitrile onto the dye sensitized anatase surface, a reorientation of the dye molecules has been deduced. In contrast, the coadsorption of unpolar benzene molecules does not involve an orientation change of the dye molecules. Thus it is concluded that the reorientation of the dye molecules depends on the polarity of the coadsorbed solvent.

The additional adsorption of electrolyte species onto the dye sensitized TiO_2 substrate, including an I^- anion, the cation (1-propyl-3-methylimidazolium) and the solvent acetonitrile, has been performed. In this work it has been postulated that by the interaction between the HOMO of the dye and the I^- ions, additional states closely below the HOMO level are formed, which are probably involved in the hole transfer from the HOMO to the reduced redox couple species, the I^- ion. By means of a homogeneous distribution of the I^- by the coadsorbed solvent, it could be shown, that the electrolyte salt is dissolved in the solvent, probably activated by the synchrotron beam. Thus the relevance of the model experiments at the TiO_2/dye/electrolyte salt/solvent interface corresponding to the real, solid/liquid TiO_2/dye/electrolyte interface is evidenced.

Übersicht

Im Rahmen dieser Arbeit wurde zum ersten Mal versucht, die komplette Arbeitselektrode (TiO_2/RuL'$_2$(NCS)$_2$/Electrolyt) der Farbstoffinjektionssolarzelle (DSSC) anhand von Photoelektronenspektroskopie zu untersuchen. Die Aufgabenstellung dieser Arbeit umfasst eher die Untersuchung der Wechselwirkungen an den Grenzflächen zwischen den Adsorbaten und dem TiO_2-Substrat als die Prozesse im Volumen des Substrates oder des Elektrolyts.

Da die Bedeckung des Farbstoffes auf dem TiO_2-Substrat nur bis zu einer Monolage beträgt, wurde die synchrotroninduzierte Photoelektronenspektroskopie als eine Methode mit sehr hoher Oberflächenempfindlichkeit angewendet, um topologische, chemische und insbesondere die elektronischen Wechselwirkungen innerhalb der Arbeitselektrode zu untersuchen. Die Photoelektronenspektren wurden überwiegend am U49/PGM-2- und TGM7-Strahlrohr bei BESSY aufgenommen, wobei auch einige Spektren am Darmstädter Integrierten System (DAISY) gemessen wurden. Für die Präparation und Analyse der Fest/Flüssig-Grenzflächen wurde ein speziell dafür entwickeltes Versuchssystem (SoLiAS) eingesetzt, welches vom Institut für Oberflächenforschung des Fachbereichs Materialwissenschaften der Technischen Universität Darmstadt betrieben wird. Alle untersuchten Grenzflächen wurden im Ultrahochvakuum (UHV) präpariert und in die Messkammer transferiert, welches eine Kontamination durch Luft ausschließt. Um die Adsorption des flüssigen Lösungsmittels im UHV durchführen zu können, wurde an der SoLiAS ein mit flüssigem Stickstoff gekühlter Manipulator eingesetzt.

Die Adsorption von Acetonitril und Benzen auf das unbehandelte TiO_2-Substrat zeigt eine Wechselwirkung der jeweiligen Lösungsmittelmoleküle mit dem Substrat unterschiedlichen Ausmaßes. Während Acetonitril unterkoordinierte Ti^{3+}-Stellen und Sauerstoffleerstellen (V_O-Zustände) in einem beträchtlichen Ausmaß dämpft, so ist die Wechselwirkung mit Benzen ziemlich schwach. Im Gegensatz zur Acetonitril-Adsorption, wo ein gewisser Anteil von Molekülen an der Substratoberfläche chemisorbiert sind, sind Benzen-Moleküle überwiegend physisorbiert.

Von besonderem Interesse für die Funktion der DSSC ist der gerichtete Transfer von Elektronen und Löchern, begründet durch die ausgeprägte Geometrie des Farbstoffmoleküls. Dieser gerichtete Transfer wird oftmals als *vektorieller Ladungstransfer* bezeichnet [1]. Das bedeutet, dass der Farbstoff als elektronische Membran fungiert, welche Elektronen und Löcher in entgegengesetzte Richtungen transferiert. Folglich ist die Orientierung des Farbstoffes und seine Bindung zum TiO_2-Substrat der entscheidende Punkt, um eine effiziente Ladungsträgertrennung zu ermöglichen. Anhand der Koadsorption polaren Acetonitrils auf die mit Farbstoff bedeckter Anatasoberfläche wurde eine Umorientierung der Farbstoffmoleküle abgeleitet. Im Gegensatz dazu bewirkt die Koadsorption unpolaren Benzens keinen Orientierungswechsel der Farbstoffmoleküle. Aufgrund dessen wird gefolgert, dass die Umorientierung der Farbstoffmoleküle von der Polarität des koadsorbierten Lösungsmittelmoleküls abhängt.

Die zusätzliche Adsorption einer Elektrolytspezies auf das mit Farbstoff bedeckten TiO_2-Substrats wurde durchgeführt, welche das I^--Anion, das Kation (1-Propyl-3-Methylimidazolium) und das Lösungsmittel Acetonitril enthält. In der vorliegenden Arbeit wurde postuliert, das durch die Wechselwirkung zwischen dem Farbstoff-HOMO und den I^--Ionen zusätzliche Zustände dicht unterhalb des HOMO-Niveaus gebildet werden, welche möglicherweise am Löchertransfer vom HOMO zur reduzierten Redoxspezies, den I^--Ionen, beteiligt sind. Anhand einer homogenen Verteilung der I^--Ionen durch das koadsorbierte Lösungsmittel konnte gezeigt werden, dass das Elektrolytsalz im Lösungsmittel gelöst wird – möglicherweise aktiviert durch den Synchrotronstrahl. Von daher wurde die Relevanz der Modellexperimente an der TiO_2/Farbstoff/Elektrolytsalz/Lösungsmittel-Grenzfläche gezeigt, welche der realen, fest/flüssigen TiO_2/Farbstoff/Elektrolyt-Grenzfläche entspricht.

Introduction to the Dye Sensitized Solar Cell

Historical background

The discovery of the photoelectric effect by the french scientist Edmond Becquerel [2] is fundamental for all subsequent activities in research and development of devices to convert light into electrical power. For his pioneering experiment he used a halide salt solution as an electrolyte between two platinum electrodes. Motivated by photography Vogel discovered that by adding dye to a silver halide emulsion [3] he could extend the sensitivity of this photoactive electrolyte to visible light. The first attempt of sensitizing a solid silver halide electrode with a dye for a photoelectrochemical cell was carried out by Moser [4] in 1887. Almost eighty years later it is generally accepted that adsorbing sensitizing dyes on semiconductor electrodes in a dense monolayer is needed for a high efficient Dye Sensitized Solar Cell (DSSC) [5]. But photocorrosion of the halide electrode occurs in the electrolyte solution. Therefore oxide semiconductors are used, because of their improved photostability. The next big step in the development of the DSSC was the change from a smooth to a porous surface to increase the adsorption of light by the dye monolayer dramatically. In 1991, O'Regan and Grätzel [6] presented a very efficient DSSC based on nanocrystalline TiO_2 showing an efficiency of 7.1% under full solar illumination and up to 12% under diffuse light. As sensitizer they used the Ru dye N3 (Ru^{II}(2,2'-bipyridine-4,4'-dicarboxylate)$_2$(NCS)$_2$), RuL'$_2$(NCS)$_2$ with L'=(2,2'-bipyridine-4,4'-dicarboxylate)$_2$ which is still among the dyes showing highest efficiencies.

Device operation

In a dye sensitized solar cell (DSSC) [6–10] the complex interplay between the TiO_2 substrate, the dye molecules and the redox couple I^-/I_3^- is crucial for the performance of the device. Unlike in a p/n junction cell the charge separation and transport take place in different media (Figure 0.1). Hence in the n-doped TiO_2 semiconductor mainly majority charge carriers (electrons) are present, whereas minority charge carriers (holes) are insignificant. Thus recombinations between electrons and holes in bulk material, as it is the case for conventional p/n junction cells, are minimized. The device operation of the DSSC can be summarized in a simplified action sequence as follows. Within the sensitizer dye molecule (S) the generation of the electron hole pair (step 1) occurs by absorption of light of sufficient energy ($h\nu \geqq \Delta E_{LUMO-HOMO}$) from the highest occupied molecular orbital (HOMO) into the lowest unoccupied molecular orbital (LUMO) [11]. Owing to photon absorption the dye molecule is in an excited state (S^*).

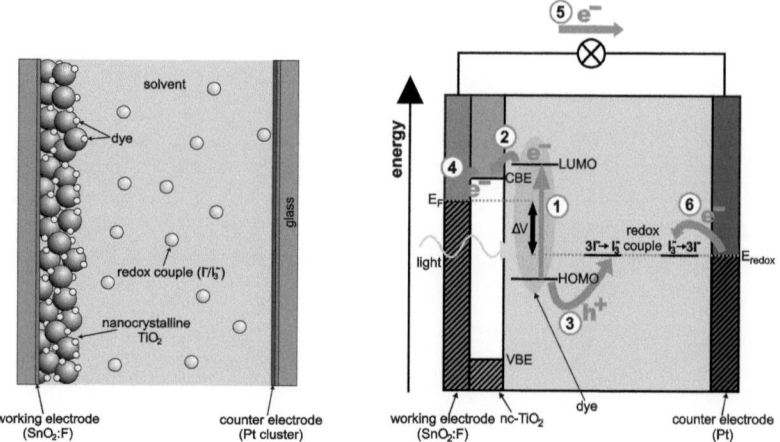

Figure 0.1: left: A schematic picture containing all relevant parts of the DSSC as indicated. **right:** The individual operation steps in the DSSC, including generation of the electron hole pair (step 1), injection of the electron into the TiO$_2$ semiconductor (step 2), reduction of the dye by I$^-$ ions forming I$_3^-$ ions (step 3), diffusion of the electrons through the nanocrystalline TiO$_2$ (step 4), electrons contributing to the current circuit (step 5), and reduction of I$_3^-$ to I$^-$ (step 6)

$$S + h\nu \longrightarrow S^* \tag{0.1}$$

The photogenerated electron is injected in a pico- to femtosecond timescale [12, 13] from the LUMO of the dye molecule to the conduction band of the nanocrystalline TiO$_2$ substrate (step2) resulting in an oxidation of the dye molecule (S^+).

$$S^* \longrightarrow S^+ + e^-(TiO_2) \tag{0.2}$$

From the HOMO level of the dye the remaining hole is captured by the reduced I$^-$ ions of the redox couple (step 3) resulting in a rereduction of the oxidized dye.

$$S^+ + e^- \longrightarrow S \tag{0.3}$$

Upon this oxidation reaction I$_3^-$ ions are formed which carry the positive charge to the counter electrode [14].

$$3I^- - 2e^- \longrightarrow I_3^- \tag{0.4}$$

It is generally assumed that within the nanocrystalline grains (10-50 nm in size) there is no space charge layer present, which results in a missing built-in field. Hence the injected electrons are transported through the TiO_2 particle film by diffusion (step 4) reaching the fluorine doped tin oxide layers (SnO_2:F or FTO), which is a transparent conducting oxide (TCO). This electrons reach the counter electrode through the external load and wiring (step 5). At the counter electrode the I_3^- ions are reduced to I^-, injecting the positive hole to the platinized counter electrode (step 6) allowing for a closed current circuit

$$I_3^- + 2e^- \longrightarrow 3I^- \tag{0.5}$$

In the equation above only the primary charge carrier species are involved, i.e. electrons in the TiO_2 film and the reduced and oxidized forms of the hole conductor (I^-/I_3^-).

Concerning the performance of the device, the following energy levels are most important (see Figure 0.1) [15]:

- the HOMO of the dye
- the LUMO of the dye
- the Fermi level of the TiO_2 substrate (E_F)
- the conduction band edge of the TiO_2 (E_{CBE})
- the redox potential of the electrolyte (E_{redox})
- the electron levels of the electrolyte (E_{red} and E_{ox})

Of essential relevance is the HOMO-LUMO energy gap ($\Delta E_{LUMO-HOMO}$) and the energy gap between the Fermi level of the TiO_2 and the redox potential of the electrolyte ($\Delta E_{E_F-E_{redox}}$). The wavelength of absorbed photons is determined by the energy gap of the dye. The maximum photovoltage is determined by and increasing with $\Delta E_{E_F-E_{redox}}$ under illumination. To assure electron-transfer reactions with optimal rate, the energy gaps $\Delta E_{E_{redox}-E_{HOMO}}$ and $\Delta E_{E_{LUMO}-E_{CBE}}$ (or $\Delta E_{E_{LUMO}-E_F}$ for a degenerated TiO_2 semiconductor) have to be sufficiently large.

Although a lot of empirical studies have been performed with the goal to increase the device efficiency, there is still a lack of comprehension in terms of the complex interplay at the interfaces of the device. This work emphasizes onto the interfacial interactions within the TiO_2/RuL'$_2$(NCS)$_2$/electrolyte salt/solvent using Synchrotron Induced Photoemission Spectroscopy, which provides information on electron states and surface (interface) potentials with highest surface sensitivity.

Part I

Fundamentals

1 The TiO$_2$/dye/electrolyte interface

This chapter summarizes all important aspects regarding the function of the working electrode (TiO$_2$/dye/electrolyte). The alignment of the electronic levels of the respective components and the formation of the electrochemical double layer, which is important for the charge transfer kinetics, will be described. At the end the chapter closes with a summary of the most important fields regarding the efforts within the scientific community how to improve the device performance. Since the scope of this work is on the TiO$_2$ – RuL'$_2$(NCS)$_2$ – I$^-$/I$_3^-$ system and the interactions within, it is essential to give an introduction into the fundamental mechanism of contact formation and electron transfer processes. Due to the completely different working mechanism compared to the conventional p/n junction cells, interfacial processes are emphasized rather than bulk processes like in conventional solar cells [16, 17].

1.1 Alignment of electronic energy levels

To ensure the function of the Dye Sensitized Solar Cell, the electronic energy levels of the respective components have to be aligned in an adequate way (Figure 1.1).

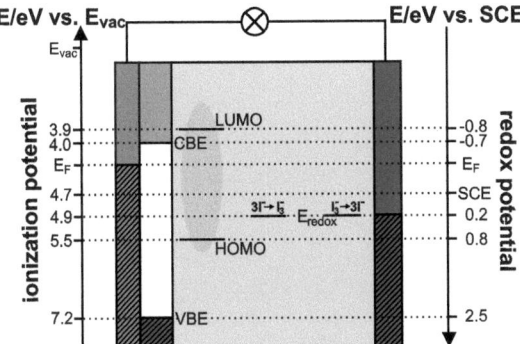

Figure 1.1: A scheme showing the electrochemical potentials of the respective components in the DSSC. For the dye values of the RuL'$_2$(NCS)$_2$ have been taken. On the right axis the electron energy levels in the different phases in the cell are given as measured electrochemically (redox potential) against the standard calomel electrode (SCE)(after [16], not drawn to scale). On the left axis these values have been converted in an energy scale against the vacuum level (ionization potential). The Fermi level depends on the doping of the SnO$_2$:F subsubstrate.

Nazeeruddin et al. [18] measured redox potentials of RuL'$_2$(NCS)$_2$ against the standard calomel electrode (SCE) with cyclic voltammetry, and a value of 0.8 eV for the oxidation potential of the HOMO level was obtained. Since with cyclic voltammetry the metal working electrode is scanned through a predefined potential window in both directions within each cycle, charge transfers are induced in both directions corresponding to an anodic and cathodic current. Since the electrochemical measurements by Nazeeruddin et al. [18] have been performed under illumination, in the initial state the HOMO level misses one electron due to photoexcitation. The situation is different in the case of Photoelectron Spectroscopy, where in the final state the HOMO level is oxidized due to photoexcitation.

The absorption threshold for RuL'$_2$(NCS)$_2$ adsorbed onto TiO$_2$ is about 780 nm [18], which corresponds to the optical gap energy of 1.6 eV, whereas the maximum of optical absorption amounts to 535 nm [18] (2.32 eV). One has to take into account, that upon optical excitation a Frenkel exciton on one dye molecule is produced. The optical gap E_{opt} between ground state (HOMO) and excited state (LUMO) of the dye is smaller by the amount of the exciton binding energy [19] compared to the transport gap E_t, where a remaining charge on the dye molecule causes a volume polarization. Compared to the gap $E_{PES/IPES}$ between HOMO and LUMO measured by Inverse Photoemission (IPES) and Photoemission Spectroscopy (PES), the polarization of both HOMO and LUMO is smaller since the molecules measured are close to or on the surface and are also affected by the surrounding solvent. The band edge potentials of TiO$_2$ (measured in aqueous electrolyte of pH = 1 [16]) and the bandgap of anatase must align to the dye in a way, that the LUMO of the dye has a lower electron affinity than the conduction band edge (CBE) of TiO$_2$, which is considered as a driving force for electron injection into the semiconductor.

A comparison with an Ultraviolet Photoemission Spectroscopy study accomplished by Snook et al. [20] shows that the ionization potential for the HOMO energy (5.47 eV vs. E_{vac}) is similar to the value determined by cyclic voltammetry (5.59 eV vs. E_{vac}). However this study proposes an ionization potential of the TiO$_2$ substrate in a way, that electron injection from the LUMO level (the onset of optical absorption) into the conduction band of TiO$_2$ seems not to be possible without overcoming some barrier.

Different values for both the dye and the TiO$_2$ ionization potentials were obtained by Synchrotron Induced Photoemission Spectroscopy investigations of Westermark et al. [21]. The HOMO level was located at an ionization potential of about 6 eV vs. E_{vac}. In addition the authors allocated the TiO$_2$ band edges at 200 meV lower ionization potentials (VBE at 7.0 eV and CBE at 3.8 eV). In contrast to Westermark et al. Liu et al. [22] found higher ionization potentials at 7.4 eV for VBE and 4.1 eV for CBE. All authors assumed the conduction band edge at the Fermi level because of the assumption, that the Fermi level is pinned at the conduction band edge [23].

One aim of this work is to investigate the alignment of the electronic levels of the components of the TiO$_2$/dye/electrolyte interface relative to each other by means of Photoelectron Spectroscopy.

1.2 The electrochemical double layer

A potential difference at the semiconductor/electrolyte interface is induced by charge separation (e.g. by electron injection, see Section 1.3.1). The semiconductor becomes negatively charged relative to the electrolyte, and cations are attracted by Coulomb interaction to build up an electrochemical double layer. The so called *Helmholtz model* [24] is the most simple model to describe this double layer (Figure 1.2 a)) as given in common textbooks on electrochemistry [25, 26].

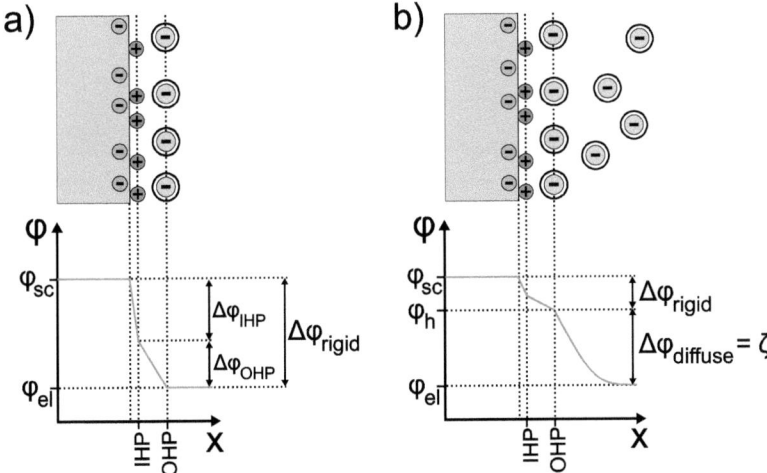

Figure 1.2: A scheme of the electrochemical potential drop across the semiconductor/electrolyte interface (adapted from [27]). The anions in the electrolyte are drawn with a solvate shell (outer circle). **a)** IHP is the inner Helmholtz plane due to strongly chemisorbed species at the surface of the semiconductor and OHP is the outer Helmholtz plane due to the counter ionic charges of solvated counter ions. **b)** The Stern model with the potential difference $\Delta\phi_{rigid}$ extending over the Helmholtz double layer and the exponential potential decay $\Delta\phi_{diffuse}$ ranging over the Gouy-Chapman double layer.

The surface plane of strongly adsorbed electrolyte components on the semiconductor is denoted as the inner Helmholtz plane (IHP). The outer plane of ionic charge in the electrolyte is called the outer Helmholtz plane (OHP). The *Gouy-Chapman model* [28, 29] takes the effect of thermal motion of the ions in the solution into account and suggests instead of a rigid double layer like in the Helmholtz model a more diffuse double layer. But neither the Helmholtz nor the Gouy-Chapman model is a sufficient description of the interface. The former overrates the rigidity of the double layer, whereas the latter underestimates its structure. Therefore the two models where combined in the *Stern model* [30] (1.2 b)), in which the ions close to the interface are constrained into a rigid Helmholtz plane and in the bulk of the solution the ions are dispersed like in the *Gouy-Chapman model*. The potential drop $\Delta\phi_{diffuse}$ from the OHP into the solution is exponential and called ζ-potential in the literature.

Applying the concept of the Stern double layer qualitatively to the TiO$_2$/dye/electrolyte interface in the DSSC has been accomplished e.g. by Zaban and Gregg [17, 31, 32] (Figure 1.3).

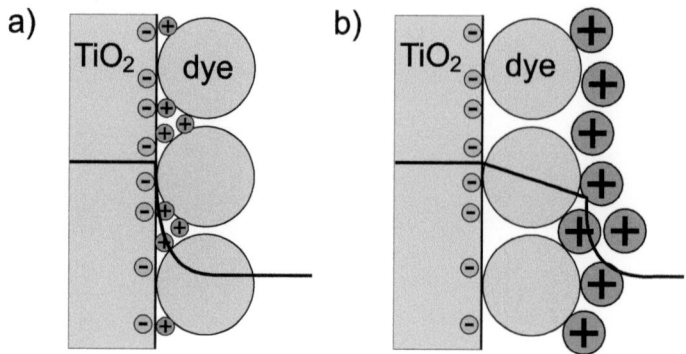

Figure 1.3: A scheme of the electrochemical potential drop from the semiconductor across the semiconductor/dye/electrolyte interface to the redox potential of the electrolyte under negative applied or photogenerated bias (from [32]). The potential drop is dependent on the cation size in electrolyte solution. **a)** In the presence of small cations (e.g. Li$^+$), the dye oxidation potential is hardly affected by changes in the TiO$_2$ potential. **b)** In the presence of large cations (e.g. 1-propyl-3-methylimidazolium iodide (PMII)) the potential of the dye follows changes of the potential of the semiconductor.

It is well known, that the conduction band potential of oxide semiconductors change 59mV per pH unit (*Nernst shift*) [33–35]. The electrochemical double layer in the interfacial region is influenced by the cation species in the electrolyte and the dye adsorbed on the TiO$_2$ surface. Using the picture of a plate type capacitor for the Helmholtz double layer, where the two Helmholtz planes represent the two plates of the capacitor, the dye lies outside in the case of small cation coadsorption on the TiO$_2$ surface (Figure 1.3 a)). Therefore the dye is hardly influenced by any potential change of the semiconductor. The opposite situation is given by coadsorption of large cations, which cannot adsorb on the TiO$_2$ surface because they cannot penetrate the dye layer due to their size (Figure 1.3 b)). Any relative change in potential of the semiconductor towards the electrolyte changes the slope of the linear electrical field within the two capacitor plates. In the latter case driving forces for all crucial interfacial reactions like photoinjection of electrons in the TiO$_2$, transfer of the hole to the counter electrode and recombination reactions of electrons with the oxidized ions in the electrolyte are determined by the semiconductor conduction band potential, the redox potential and on the spatial extend of the potential at the interfacial region (double layer).

In this work, by coadsorption experiments of dye, electrolyte salt and solvent, the electronic alignment within the working electrode half cell needs to be elucidated. The size of the

1.3 Charge transfer processes

electrochemical double layer can be varied by adsorbing electrolyte salts with different cation sizes.

1.3 Charge transfer processes

The various electron transfer processes in the Dye Sensitized Solar Cell differ in orders of magnitude with respect to their kinetics. Figure 1.4 shows that the desired electron transfers have much higher rate constants than the unwanted recombination reactions, which is considered to be the reason for an effective charge separation of the DSSC.

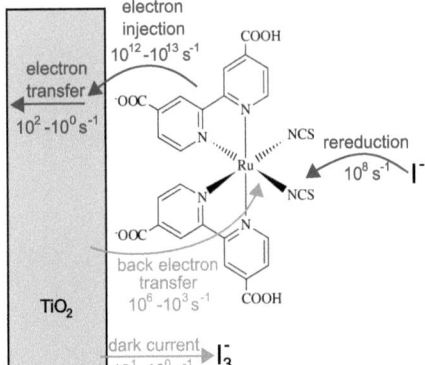

Figure 1.4: Schematic diagram of electron transfer processes. The arrows pointing left (blue) represent desired electron transfers, whereas the arrows pointing right (red) represent unwanted recombination processes (adapted from [15], values taken from [36]).

The kinetics of all charge transfer reactions depend on the alignment of the electronic energy levels of substrate, dye and electrolyte. By this present work it is attempted to investigate the position of all relevant energy levels of the involved components. Photoelectron Spectroscopy is an appropriate method for this purpose.

1.3.1 Electron injection

As one of the fastest electron transfer processes observed in nature the electron injection of photoexcited electrons in the conduction band of the TiO_2 occurs within pico- to femtoseconds (<100 fs) [12,13]. The rate constant for electron injection k_{inj} can be described by Fermi's golden rule:

$$k_{inj} = \left(\frac{4\pi^2}{h}\right) |M_{fi}|^2 \rho(E) \tag{1.1}$$

where $M_{fi} = \langle \psi_f | V | \psi_i \rangle$ is the matrix element, which describes the transition from an initial state $| \psi_i \rangle$ to a final state $\langle \psi_f |$ due to the perturbation V. In this case $\rho(E)$

is the density of states in the conduction band and h the Planck constant. Overall the electron injection rate k_{inj} depends on the overlap of the wavefunctions of the excited state in the dye (LUMO) and the conduction band states of equivalent energy position in the TiO_2. Hence this transfer is ultrafast due to a strong coupling (bonding interaction) of the adsorbed $RuL'_2(NCS)_2$ dye to the TiO_2 substrate (see Section 2.2.2). Of course the injection process strongly depends on the semiconductor with respect to the energetic alignment of the conduction band edge to the LUMO of the dye. The large density of states in the conduction band of TiO_2 favours an ultrafast electron transfer. The conduction band edge energy can be lowered and therefore the injection rate can be enhanced by adsorption of cations on the surface [37,38]. Unfortunately a downward shift of the conduction band edge enhances not only the injection rate, but also recombination reactions. A biphasic electron injection process was discussed by Anderson et. al [13], which consists of two electron injection paths with different rate constants. Besides the ultrafast electron injection from unrelaxed excited states in the dye another transfer process from thermalized excited states of the dye into the oxide was suggested, which is much slower and therefore not relevant. The injection time of $RuL'_2(NCS)_2$ on different semiconductor substrates (SnO_2, ZnO and In_2O_3) has been compared with the injection time on TiO_2 [39]. Due to the highest density of acceptor states for the photoinjected electrons the injection process appeared to be the fastest on TiO_2. Therefore TiO_2 is the best option today for sensitization with Ru complex derived dyes.

1.3.2 Dye rereduction

Due to the oxidation of the dye by the electron injection process the rereduction of the dye is necessary. Therefore the redox couple I^-/I_3^- is employed and without it the dye would just been rereduced by recombination with the injected electron. The rereduction of the dye (Equation 0.3 and 0.4) can be summarized by the following sum reaction:

$$S^+ + \frac{3}{2}I^- \longrightarrow S + \frac{1}{2}I_3^- \qquad (1.2)$$

It can be divided into the following detailed reaction steps [40]:

$$S^+ + I^- \longrightarrow S + I\bullet \qquad (1.3)$$
$$I\bullet + I^- \longrightarrow I_2\bullet^- \qquad (1.4)$$
$$2I_2\bullet^- \longrightarrow I_3^- + I^- \qquad (1.5)$$
$$I_2\bullet^- + e^- \longrightarrow 2I^- \qquad (1.6)$$

Transient absorption spectra of the decay of the oxidized state of the $RuL'_2(NCS)_2$ dye conducted by Montanari et al. [41] show a dependence of the yield of $I_2\bullet^-$ species on the concentration of the employed redox couple. At lower concentration of the redox

1.3 Charge transfer processes

couple (< 30 mM) there is a kinetic competition between the rereduction of the dye by the redox couple and the injected electrons in the semiconductor. Employing higher electrolyte concentrations leads to a predominance of the rereduction by the redox couple. Also the amount of dye coverage on the TiO_2 surface is important. It was reported that at higher coverages than 0.3 monolayer of dye the incident photon-to-current efficiency (IPCE) and the absorbed photon-to-current efficiency (APCE) showed an abrupt increase [42]. A hopping mechanism of holes between the dye molecules has been suggested to explain this behavior. The rereduction of the dye is crucial for the cell performance, since the oxidized dye has a lifetime of up to 1 second before it decomposes. But due to the ultrafast reduction one dye molecule can withstand up to 10^8 redox cycles.

The electron injection and the dye rereduction accomplish the charge separation in the Dye Sensitized Solar Cell. The general nature of this charge separation is given by the geometry of the dye and is denoted in literature as a *vectorial charge transfer* [1], which is explained in Section 2.2.2. The adsorption geometry of the dye on the substrate and Fermi's Golden rule are connected: As long as the dye is adsorbed in the favorable geometry, the matrix element of initial and final state is large (Equation 1.3.1).

Therefore the orientation of the dye relative to the substrate surface is of prime interest and needs to be elucidated within this present work. Since the dye has a distinct and asymmetric geometry, it is possible to detect the adsorption arrangement of the dye onto the substrate surface by SXPS experiments. This is possible by means of the respective intensities of the photoemission lines of the atoms within the dye molecule.

1.3.3 Electron transport

The electron transport in TiO_2 has attracted a vast interest in the scientific community [43–54] because of its importance for the device performance of the DSSC. Despite of that a comprehensive model for electron transport in the nanoporous TiO_2 is still not established. Since in the semiconductor/electrolyte interface the charge density in the electrolyte is much larger than the charge density in the semiconductor, this semiconductor/electrolyte interface is considered being analogous to the semiconductor/metal contact. Thus in a conventional semiconductor/electrolyte contact with a flat interface [55] the potential difference is accommodated by band bending in the semiconductor forming a space charge layer.

But in the DSSC containing a nanoporous semiconductor, the situation could be completely different, if the following two preconditions are given: First each particle of the highly porous nanocrystalline electrode has contact to the ambient electrolyte (Fig. 1.5). Second the TiO_2 semiconductor is intrinsic or only lightly doped. If these two preconditions match, no substantial potential drop occurs within the nanoparticles [1, 16]. The high porosity of the TiO_2 electrode prevents formation of a space charge layer across the thickness of the semiconductor or even across several particles due to screening by the ambient electrolyte. Since lightly doped TiO_2 nanoparticles are too small to accommodate a substantial potential difference, a space charge layer within the semiconductor cannot be formed [56]. According to calculations the nanoparticles are only able to accommodate a potential drop of 50 meV between the surface and the center of a 10 nm particle radius of (lightly n-type) TiO_2 [57].

In contrast, our Synchrotron Induced Photoemission Spectroscopy experiments indicate that the Fermi level of the TiO_2 substrate is in the conduction band, thus the substrate is highly n-doped (Fig. 6.10 right). Also the morphology of the nanoporous TiO_2 does not seem to be so porous that each particle has contact to the electrolyte, as indicated by Scanning Electron Microscope images (Fig. 6.3, page 80 and Fig. 6.5, page 82).

1.3.3.1 Screening

Because of the nanocrystalline morphology of the TiO_2 substrate, the liquid electrolyte penetrates the "bulk" of the oxide. This interpenetration of two bicontinuous phases has a multitude of consequences. It neutralizes all electric fields, equilibrium or photoinduced [17] within 1 nm throughout the bulk and 15 nm at the substrate surface. The TiO_2 semiconductor and the electrolyte have different conductivities and polarizabilities, and for the semiconductor they are highly dependent on external influences like light intensity and energy. Hence the path of the photoinjected electrons varies with the illumination of the DSSC, which is described by a simple distributed resistor model in [58].

In addition to the macroscopic electrical screening by the interpenetration of bicontinuous phases as explained above, there occurs screening of electrons in the semiconductor by cations in the electrolyte (Figure 1.5).

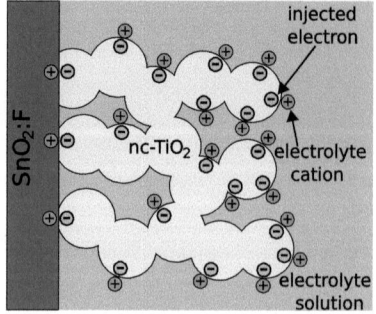

Figure 1.5: The screening of photoinjected electrons in the nanocrystalline TiO_2 by the cations in the electrolyte solution implies, that the electrons reside close to the surface of the oxide. Therefore surface states are crucial for the conduction mechanism of electrons in the TiO_2. Also electrons in the TiO_2 are compensated by opposite charges in the SnO_2:F layer (adapted from [59]).

Mobile cations in the electrolyte and electrons in the semiconductor are attracted by Coulomb interactions to each other. Thus electrons are located close to the surface of the nanoporous TiO_2. The potential drop between electrolyte and semiconductor occurs at the Helmholtz double layer (see Section 1.2).

1.3 Charge transfer processes

1.3.3.2 Ambipolar diffusion

Due to the absence of a built-in electrical field inside the electron conducting TiO_2 the electrons are moving through the nanoporous material by ambipolar diffusion [60–62]. This concept takes the mobility of both electrons and cations into account. The screen charge of the electrons, i.e. the cations in the electrolyte, move together with the electrons in the semiconductor as a neutral quasiparticle across the TiO_2 surface. As a consequence the role of surface states may be important for the conduction mechanism in the nanocrystalline anatase substrate. Electrons and ions are separated on a distance smaller than the size of a nanocrystalline TiO_2 particle [48]. A simplified ambipolar diffusion coefficient D_{amb} describes quantitatively the transport of charges in the nanoporous semiconductor/electrolyte system [48, 61, 63]:

$$D_{amb} = \frac{n+p}{n/D_p + p/D_n} \approx D_n \qquad (1.7)$$

where D_p and D_n stand for the respective diffusion coefficients of cations and electrons and p and n represent the concentration of cations and electrons respectively. Due to the much higher ion concentration in the electrolyte ($p \sim 10^{21}$ cm^{-3}) compared to the electron concentration in the TiO_2 ($n \sim 10^{18}$ cm^{-3}), the ambipolar diffusion coefficient D_{amb} is approximately the same as the electron diffusion coefficient D_n.

1.3.3.3 Percolation

However electron transport is strongly influenced by the film morphology of the TiO_2 semiconductor. The area and number of the interconnections of an anatase nanoparticle to other particles play a keyrole in electron transport dynamics. It was found by modelling a string of particle spheres, that the smaller the area of interconnections the slower is the electron transport through the material [50]. The coordination number of individual particles, i.e. the average number of interconnections, is determined by the film porosity. It was found that the average coordination number ranges from 2.2 at 80% porosity to about 4.5 at 50% porosity [64]. There is a power law dependency on the diffusion coefficient of electrons in the nanoporous material D_n of the difference between a critical porosity[1] P_c and the film porosity P:

$$D_n \propto |P_c - P|^\mu \qquad (1.8)$$

with μ as the conductivity exponent (0.82 ± 0.05). The critical porosity was found to be 0.76 ± 0.01 [64]. Additionally the length of the electron path is increasing by increasing porosity. In summary, electron transport is enhanced with decreasing porosity of the film.

---
[1] Below the critical porosity P_c there is at least one way through the entire material from one side to the other side [65].

1.3.3.4 Trapping/detrapping

The electron transport depends strongly on the intensity of incident light, which is due to a broad distribution of traps in the semiconductor [66]. This intensity dependence of the electron diffusion coefficient D_n reflects the change in the ratio of free to trapped electrons, while the quasi Fermi level rises up towards the conduction band. Therefore electron transport can be explained in terms of a multiple trapping model [44, 46, 67], where the ambipolar diffusion coefficient is rising exponentially with the photogenerated electron density [48, 52, 61], which was determined by measurements of the photocharge generated by a laser pulse (transient photocharge measurements):

$$D_n \propto n^{(1-\alpha)/\alpha} \tag{1.9}$$

where n is the photogenerated electron density and $\alpha = k_B T/m_c$ (with $0 < \alpha < 1$) is the dispersion parameter with m_c as the average trap depth. To account for the exponential increase of the diffusion coefficient with the photogenerated electron density, an exponential energy distribution of the trap states $g(E)$ was deduced [52, 61, 68–70]:

$$g(E) = g_0 exp\left(\alpha \frac{E - E_c}{k_B T}\right) \tag{1.10}$$

with E as the trap energy, E_c the conduction band edge energy and g_0 as a constant. The equivalence of the multiple trapping model to the random walk approach has been shown [49]. Due to thermalization of electrons they are trapped in energy distributed trap states and upon light illumination detrapped to the conduction band again (see Figure 1.6).

The driving force for the transport of electrons through the material from the dye sensitized surface to the SnO_2:F electron collecting electrode, which is the prerequisite for the diffusion mechanism, is the concentration gradient of electrons inside the material. It might be possible, that the concentration gradient results in a buildup of an electrical field across the TiO_2 layer [72]. A direct evidence for the trapping of electrons in trap states in the nanoporous TiO_2 upon light illumination comes from measurements in which charge is extracted in the dark [73, 74]. The spatial distribution of these trap states has been investigated, and although it has not been proven directly they have been attributed as surface states [75]. However, the role of these surface states for the device performance is still unclear. Some groups state that the trap relaxation is much faster than the rate for the charge transfer to I_3^- ions [54, 71, 76]. This is in contradiction to the findings of other groups, which consider the recombination with I_3^- as the predominant process compared to trap relaxation [53, 77]. But a common assertion of these groups is that recombination and trapping of electrons compete with each other.

In contrast to the attribution of surface states being possible recombination centers [76, 78], a beneficial effect of surface states for the conduction of electrons in the TiO_2 semiconductor was reported [79, 80]. It was observed that upon UV light illumination the short

1.3 Charge transfer processes

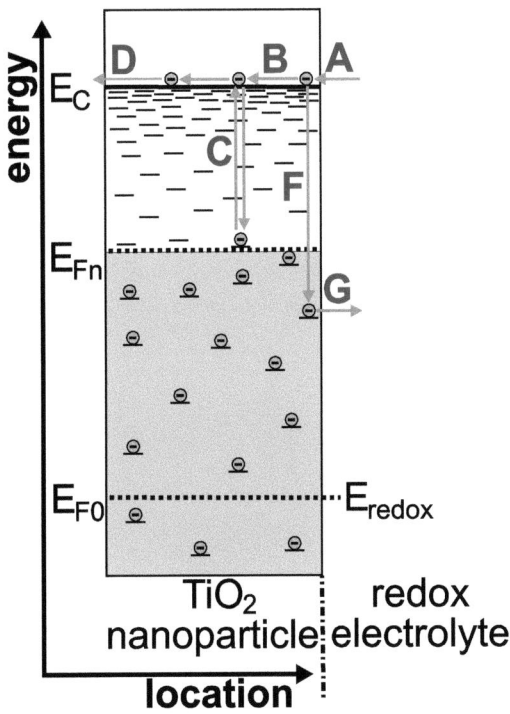

Figure 1.6: The trapping/detrapping and recombination mechanisms of electrons inside the nanoporous semiconductor over one nanoparticle (adapted from [71]). The Fermi level in the dark is denoted with E_{F0}, whereas under light illumination E_{Fn} is the quasi Fermi level. The driving force for diffusion is the concentration decay of electrons towards the SnO$_2$:F electrode. E_C stands for the conduction band minimum of the semiconductor and E_{redox} for the redox potential of the electrolyte. **(A)** Electron photoinjection from excited dye molecules; **(B)** Electron transport through delocalized conduction band states; **(C)** Electron capture and release at an exponential distribution of localized trap states; **(D)** Electron transfer to an adjacent nanoparticle; **(F)** Electron capture by surface states; **(G)** Electron back reaction through surface states.

circuit current density increases dramatically (up to 45 times). Most likely this is due to a reversible production of surface state density in the TiO$_2$ semiconductor. It was interpreted in a way, that a large enough density of surface states causes a motion of electrons through sub-bandgap states below the conduction band [80]. This UV-induced effect occurs only in the presence of non-adsorbing big cations like TBA$^+$ and is inhibited by small adsorbing cations like Li$^+$ due to a surface modification of the TiO$_2$ [17].

Although trapping/detrapping is considered being the predominant transport mechanism in the device operation of the DSSC, there occurs nonthermalized electron transport (hot electrons) as well when light enters the cell through the substrate (normal device operation) [52]. A fraction of electrons are collected at the electrode before they can thermalize with traps inside the oxide. Nevertheless overall the transport in the DSSC can be described by trap limited ambipolar diffusion on percolation paths.

An aim of this Photoelectron Spectroscopy study is to investigate, how these trap states can be influenced. Since it is assumed, that these trap states arise also from oxygen vacancy states at the substrate surface (Section 2.1.4), the behavior of these traps have to be investigated in the presence of all involved adsorbates at the working electrode.

1.3.4 Charge recombination processes

Recombination processes in the Dye Sensitized Solar Cell are related to reactions of photoinjected electrons in the nanoporous semiconductor with oxidized ions, located in the electrolyte or oxidized dye molecules at the TiO$_2$ surface. In addition collected electrons in the SnO$_2$:F subsubstrate can recombine with oxidized electrolyte ions, if the SnO$_2$:F surface is not covered completely with a dense layer of TiO$_2$. Charge recombinations are loss processes, the subject of ongoing research is to find ways to minimize them as much as possible [81].

1.3.4.1 Dark current

Since the reaction, which contributes for the dark current, is of second order, it is a very slow reaction with a rate constant of second magnitude. The net reaction is described by

$$I_3^- + 2e^- \text{ (TiO}_2\text{)} \longrightarrow 3I^- \tag{1.11}$$

and involves the detailed reaction steps

$$I_3^- \longrightarrow I_2 + I^- \tag{1.12}$$

$$I_2 + e^- \longrightarrow I_2\bullet^- \tag{1.13}$$

The $2I_2\bullet^-$ radical can react in different ways, depending on whether the radical reacts with another same radical [76, 82]

$$2I_2\bullet^- \xrightarrow{slow} I^- + I_3^- \qquad (1.14)$$

or takes up an additional electron from the TiO_2 [68, 83]

$$I_2\bullet^- + e^- \xrightarrow{slow} 2I^- \qquad (1.15)$$

Despite of its slow kinetics these electron transfer processes are responsible for the dark current of the Dye Sensitized Solar Cell.

1.3.4.2 Back electron transfer (BET)

In contrast to the electron injection into the TiO_2 the much slower recombination of injected electrons with oxidized dye cations occur on the order of micro- to milliseconds. Nevertheless this process is much faster compared to the dark current. The so called back electron transfer (BET) [84, 85] is much slower than the electron injection due to the adsorption geometry of the molecule on the substrate, which is described in detail in Section 2.2.2. HOMO and LUMO of the dye are located on different spatial positions on the molecule (see Section 2.2.1). Unlike the LUMO the HOMO is pointing away from the TiO_2 substrate. Hence electrons must tunnel a much longer distance for the back electron transfer compared to electron injection. Overall the BET is even considered being negligible in a normal DSSC employing the I^-/I_3^- redox couple [77, 86] due to a very fast rereduction of the dye by the redox couple.

In order to confirm, that the BET does not affect the devices performance significantly, the favorable orientation of the dye molecule onto the substrate needs to be evidenced by Synchrotron Induced Photoemission Spectroscopy.

1.4 Ongoing development of the DSSC

To get the Dye Sensitized Solar Cell to a level suitable for a productive and long term stable solar energy conversion cell, a couple of problems must be solved. The most important problem is also the most obvious one concerning the manageability of a liquid solar cell. Since the DSSCs, which reach high efficiencies, employ a liquid electrolyte, this hole conductor dries out easily due to the high vapor pressure of commonly used solvents (e.g. acetonitrile). Many attempts have been undertaken to develop a solid hole conductor [87–91] or a polymer gel electrolyte [92, 93], but the efficiencies are much lower than in the case of a liquid electrolyte. Of course the charge transport is much lower in a polymer gel and a solid state hole conductor. But also the interpenetration of semiconductor and electrolyte is strongly reduced [94], which affects the functionality of the cell.

Also the interfacial recombination kinetics and attempts of keeping it as low as possible is subject of ongoing research [81]. The TiO_2 substrate was coated with a metal oxide blocking layer of Al_2O_3 to minimize recombination reactions [95].

Another interesting approach to improve the cells performance is to decrease the transport distance of electrons throughout the TiO_2 towards the electron collecting SnO_2:F subsubstrate. This is done by extending the SnO_2:F into the nanoporous material [96].

Of course, these few examples of active research areas do not cover the entire huge field of the ongoing development of the DSSC. There are a lot more features of the DSSC to be improved. But overall, since the idea of sensitizing nanoporous TiO_2 in 1991 by O'Reagan and Grätzel [6], which improved the cell efficiency enormously compared to flat substrates, a real breakthrough is missing. A critical assessment by Tributsch of the learning curve of the DSSC compared to classical crystalline and thin film solar cells reveals a much slower progress in improving the efficiency of DSSCs with time [97]. The author's feeling about the future of the DSSC is that only small empirical steps like in the development of photography will be possible due to the large number of interdepending parameters affecting the overall cell performance.

In this work one goal is to investigate the adsorption geometry and bonding of the dye depending on its environment. As the vectorial charge transfer is necessary to provide a fast charge separation, the spatial arrangement of the electronic states of the dye depending on the dye molecule orientation is crucial for the performance of the DSSC.

2 The investigated components of the Dye Sensitized Solar Cell

In this chapter the respective components of the Dye Sensitized Solar Cell are described in detail with respect to their material properties and relevance in device operation. Beginning with the substrate, the thermodynamic stability of the different polytypes of TiO_2 and their surfaces are evaluated, since the crystalline phases have different electron mobilities, which is crucial for the performance of the DSSC. A description of the nanocrystalline morphology of anatase as the most thermodynamic stable phase is given. Also the electronic and crystalline structure of the surfaces are presented. Furthermore surface defects on anatase are reported, because they are addressed to be a possible recombination path in the cell. The sensitizing N3 dye is described with respect to its molecular and electronic structure, emphasizing the distinct electronic and spatial configuration of the HOMO and LUMO level. Thus its adsorption geometry and anchoring mechanism turns out to be very important. A detailed report on the electrolyte and its components is given focusing on solvation properties, because they change the electronic energy levels by reorganization enthalpy terms. Being essential for the solvent–substrate interface the adsorption mechanisms of the solvent molecules are described.

2.1 The TiO_2 substrate

Titanium dioxide is one of the most widely investigated single crystal system within the field of metal oxide surface science [98]. This is because of its vast variety of applications: Besides its utilization as an electron conductor in the DSSC, as described in this work, it is also used in heterogeneous catalysis, as a photocatalyst (e.g. for hydrogen production), as a gas sensor, as a pigment in wall paint and cosmetics and as a coating material for protection against corrosion and for optical applications. Furthermore it is discussed as a material for gate insulators, as a spacer material in magnetic spin-valve systems, for lithium based batteries, for electrochromic devices and it plays an important role for the biocompatibility of bone implants [98].

2.1.1 Crystal phases and surfaces

Titanium dioxide crystallizes in three major crystal structures at normal pressure (polymorphs): rutile, anatase and brookite, but only rutile and anatase play an important role in the applications of TiO_2 (Figure 2.1). In 1916 the crystal structure of anatase and rutile were first described by Vegard [99] and later redetermined in order to get structure data

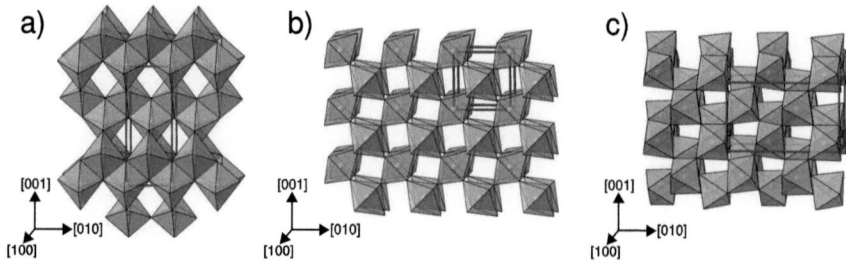

Figure 2.1: The crystal structures for the three TiO$_2$ polytypes with unit cell shown (drawn to scale, values taken from [100]). The basic buildup units are distorted [TiO$_6$]$^{2-}$ octrahedra, where titanium atoms are coordinated by 6 oxygen atoms. These octrahedras share both corners and edges in all three structures. **a)** The anatase structure with the unit cell dimensions of $a = b = 3.784$ Å, $c = 9.515$ Å (the lengths of the unit cell a, b and c are in [100], [010] and [001] direction respectively). **b)** The dimensions for the unit cells are for rutile $a = b = 4.594$ Å, $c = 2.959$ Å. **c)** For the orthorhombic structure brookite the unit cell measures $a = 9.184$ Å, $b = 5.447$ Å and $c = 5.145$ Å.

with higher accuracy [100]. Both rutile and anatase have tetragonal symmetry, but different space groups (anatase: I4$_1$/amd, rutile: P4$_2$/mnm). Unlike anatase and rutile, brookite has orthorhombic symmetry (space group Pbca).

The average Ti–O bond length rises in the order rutile < brookite < anatase, which makes rutile the most and anatase the least dense TiO$_2$ crystal structure. The arrangement of the [TiO$_6$]$^{2-}$ building units in the anatase structure leaves open channels parallel to the c-axis, which promotes intercalation e.g. of small cations like Li$^+$ [101–103]. Thus the intercalation of lithium cations into the TiO$_2$ substrate is faster for the (001) surface compared to the (101) surface, because of the smaller surface density and open channels along the crystal c–axis.

The crystal modification anatase has a higher electron mobility ($\mu = 10\,\mathrm{cm}^2/(\mathrm{Vs})$) compared to rutile ($\mu = 1\,\mathrm{cm}^2/(\mathrm{Vs})$) [98, 101], which makes anatase superior for Dye Sensitized Solar Cells. TiO$_2$ is a semiconductor having a wide direct bandgap (anatase: $E_g = 3.2\,\mathrm{eV}$ [101, 104] and rutile: $E_g = 3.0\,\mathrm{eV}$ [101, 105]), thus being transparent for visible light for a defect free crystal.

The anatase crystal structure is shown in detail in Figure 2.2. It may be derived from the NaCl prototype lattice by removing half of the Ti atoms in order to get the stoichiometry of TiO$_2$.

The formation energy for the crystal surfaces (Table 2.1) were calculated by Lazzeri et al. for anatase [107] and by Ramamoorthy et al. for rutile [108] according to the Local Density Approximation (LDA).

For anatase the (101) surface is the thermodynamically most stable surface followed by the (100) and (001) surface. Although the (100) surface should be quite stable, it does

2.1 The TiO₂ substrate

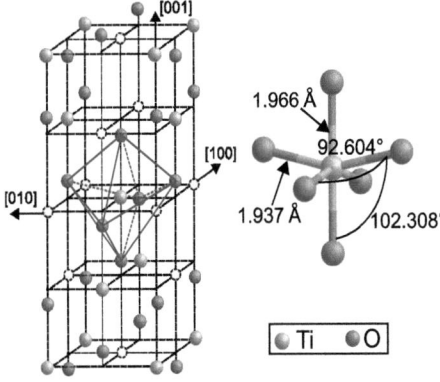

Figure 2.2: left: The anatase unit cell derived from the NaCl lattice [106]. The oxygen atoms are displaced slightly causing the distortion of the $[TiO_6]^{2-}$ octahedra. The dotted spheres represent removed metal atoms from the NaCl structure. **right:** The $[TiO_6]^{2-}$ showing the Ti–O distance and the angles between the bonds [98]. A clear deviation from a 90° bond angle exists. The Ti–O bond length varies: 2 bonds are slightly longer than the remaining four in the coordination octahedron.

	Anatase			Rutile			
surface [hkl]	(101)	(100)	(001)	(110)	(100)	(101)	(001)
formation energy [J/m₂]	0.84	0.96	1.38	0.84	1.06	1.31	1.56

Table 2.1: The surface formation energies in J/m₂ for relaxed, unreconstructed TiO₂ surfaces according to the Local Density Approximation (LDA). The values for rutile are converted from [108] and for anatase are taken from [107].

not appear in crystals (Figure 2.3 a)), but is observed in powder materials. It has been shown that the (100) surface undergoes a $(1 \times n)$ reconstruction [109], which is explained by microfaceting forming (101) surfaces [110]. The (001) surface shows a (1×4) reconstruction [111].

For rutile the (110) surface is the most stable surface, followed by (100), (101) and (001). As being comparably unstable the (001) surface does not exist in the equilibrium shape of rutile (Figure 2.3 b)). The equilibrium shapes resemble closely the shape of minerals of anatase and rutile found in nature (Figure 2.3 c) and d)).

Experimental evidence of the calculations described above has been provided. Indeed, the (101) surface is the most frequently exposed surface followed by the (100) and (001) surface for anatase nanoparticles according to Transmission Electron Microscopy (TEM) Measurements [112].

The (101) anatase surface constitutes more than 94 % of the entire anatase crystal surface compared to the presence of about 54 % of the (110) surface of the rutile crystal surface. The average surface energy for a macroscopic crystal consists of the sum of the energies of the exposed surfaces weighted with the corresponding surface area as given in the Wulff construction. Thus the average surface energy for anatase (0.90 J/m₂) is about 20% lower than for rutile crystals (1.09 J/m₂) as follows from LDA calculations [107]. One can deduce that especially for small crystal sizes like in nanopowders, where the surface energies play a big role compared to bulk properties, nanoparticles are more stable being in the anatase phase.

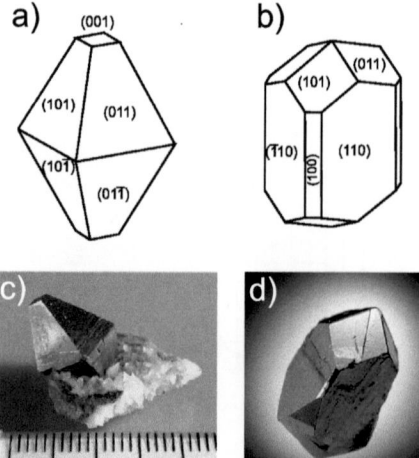

Figure 2.3: The equilibrium shapes of both anatase and rutile according to the Wulff construction (with values of Table 2.1) compared to minerals found in nature. **a)** In the anatase crystals, only the two surfaces (101) and (001) are exposed to ambience [107]. **b)** Rutile crystals show more surfaces, namely the (110), (100) and (011) surface [108]. **c)** Photograph of an anatase mineral [98] and **d)** rutile mineral [113].

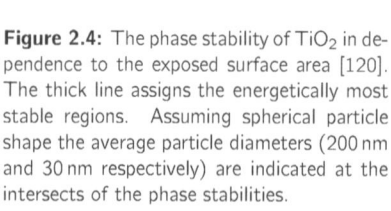

This conclusion is supported by several studies about the phase stability being a function of the particle size by experiments [114] and theory [115, 116]. Zhang et al. found anatase being stable below particle diameters of 14 nm, whereas Barnard et al. calculated values of 2.6 nm. In addition calculations [117] and experimental results obtained by calorimetric measurements of the phase stability show the occurrence of the three polytypes in dependence of the surface area [118–121](Fig.2.4). Not only the surface area (and therefore the particle size), but also impurities on the surface influence the phase stabilities of the titanium dioxide polymorphs. Barnard et al. calculated, that for a complete surface hydrogenation of TiO_2 the anatase phase is stable up to particle diameters of 23.1 nm [115, 116].

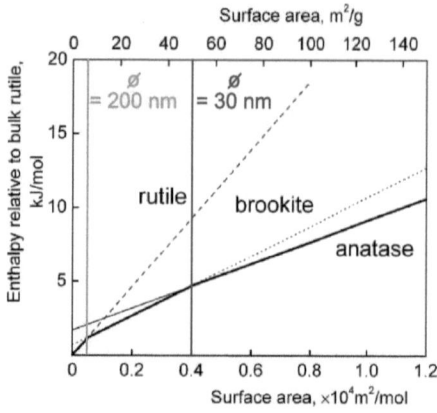

Figure 2.4: The phase stability of TiO_2 in dependence to the exposed surface area [120]. The thick line assigns the energetically most stable regions. Assuming spherical particle shape the average particle diameters (200 nm and 30 nm respectively) are indicated at the intersects of the phase stabilities.

Like shown in the graph, the anatase phase is energetically more favorable than rutile and brookite with rising surface area. All in all one can conclude that the convenient conditions

2.1 The TiO$_2$ substrate

of anatase phase stability for nanoparticles combined with the superior bulk properties of the anatase phase are important factors of employing anatase nanopowders in the DSSC.

Due to the fact, that the (101) surface is the most prevalent surface for anatase, in the following only this surface will be considered. This surface looks very corrugated like a sawtooth profile (Fig.2.5) and consists of O–Ti–O double chains along the [010] direction.

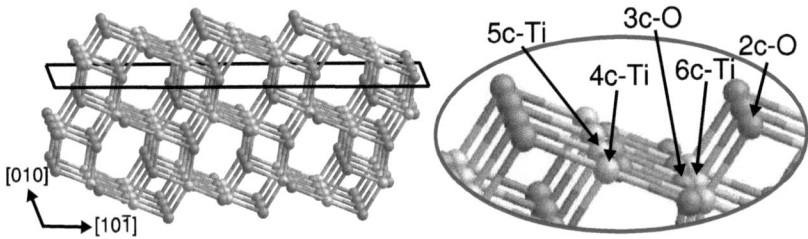

Figure 2.5: The anatase (101) surface (after [122]). **left:** Side view indicating the (101) plane (rectangle). **right:** Magnified area viewing titanium and oxygen atoms with different coordination numbers. For a detailed explanation see text.

Within these chains the titanium atoms are in fivefold (5c) or sixfold (6c) and the oxygen in threefold (3c) coordination, whereas these chains are connected by twofold (2c) coordinated oxygen atoms. 50 % of 5c and 6c Ti atoms and 50 % of 2c and 3c oxygen atoms, respectively, are present at the (101) surface. Lower coordinated Ti atoms (fourfold coordination (4c)) reside at step edges and it was found by STM measurements, that they are preferred adsorption sites [109, 122, 123]. Upon relaxation the 2c oxygen and 5c titanium atoms relax 0.06 Å and 0.17 Å into bulk direction, whereas the 3c oxygen and the 6c titanium atoms move 0.21 Å and 0.11 Å out of the surface [124]. In a detailed report of steps on anatase (101) [122] it was found that the reactivity of the local structure of the step is similar to the reactivity of the corresponding extended surface. In some cases this has the astonishing consequence that the step reactivity is lower than that of the flat surface.

This model of the coordination of titanium and oxygen atoms holds not only for an ideal (101) single crystalline anatase surface, but also for nanocrystalline anatase. There is a correlation between the oxidation state and coordination of an atom. An undercoordinated titanium atom (5c-Ti and lower coordinated Ti) is in a more reduced state relative to the fully coordinated 6c-Ti, which is Ti^{4+}. The equivalent is the undercoordinated oxygen (2c-O), which is in a more oxidized state compared to the fully coordinated 3c-O.

2.1.2 Shape of nanocrystalline crystallites

As already mentioned in the introduction (page 9) the anatase particles must be nanocrystalline in their application in the Dye Sensitized Solar Cell. The nanocrystallinity of the anatase particles ensures a large dye uptake due to a large surface area. As a consequence the light absorption by the dye sensitized TiO$_2$ electrode increases by several orders of magnitude compared to a flat dye sensitized single crystal surface. The anatase nanoparticles are of spherical shape as shown in Fig.2.6.

Figure 2.6: An SEM micrograph of nanocrystalline TiO$_2$ which was made by sintering a colloidal paste from Solaronix SA at 450° for 30 min (from [125]). The particle size ranges from 30 - 40 nm.

The particle size and the interconnection of the spherical particles are increased with the calcination temperature [126, 127]. The dark resistivity of nanoporous anatase (10^8 - 10^{10} Ωcm) [40] is much higher than the one of rutile and anatase single crystals ($\sim 10^3$ Ωcm) [128] due to enhanced electron scattering and potential barriers at grain boundaries. Hence the conductivity is increased with the particle size [129]. Also the enormous surface area of TiO$_2$ nanoparticles implies a high density of surface defects, which changes the conductivity. Although defects are usually responsible for a decrease of conductivity in conventional solar cells, in the DSSC this seems to be not the case.

2.1.3 Electronic structure

The Ti–O bonding, which mostly involves O 2p and Ti 3d atomic orbitals, has a strong ionic character. Thus the partial density of states are different for valence band and conduction band involving different contributions of the O an Ti atomic orbitals.

In the valence band spectrum measured with Ultraviolet Photoemission Spectroscopy (Fig.2.7 a) at top) one can see a strong emission, which is attributed predominantly to the O 2p orbitals (BE from around 10 eV to 3.6 eV). This emission can be divided into a nonbonding (BE = 5.4 eV) and a bonding state (BE = 8.2 eV) [131, 132]. In the bandgap a comparably weak Ti 3d contribution is located at a binding energy of around 1 eV (Fig.2.7 a) top)). The calculated partial density of states (pDOS) (Fig.2.7 a) at bottom) display the respective contribution of the atomic orbitals to the molecular orbital structure. In the valence band the pDOS amounts 75 % for O 2p states and 25 % for Ti 3d states, whereas the conduction band is mainly of Ti 3d nature (90 %), and the contribution of O 2p states is small with 10 %. This pDOS calculations matches quite well with the experimental data. Due to a shortcoming of the LDA to underestimate the band gap of the calculated material it turns out to be only 2 eV instead of the literature value of 3.23 eV. A more recent calculation [133] takes the on-site Coulomb interaction due to the strong localization of the d-orbitals into account and results in a band gap value of 3.18 eV, which is almost the same as the experimental value [101, 104]. Overall, according to the ionic character of TiO$_2$ the valence band is derived mainly from O 2p states and the conduction band originates mainly from

2.1 The TiO$_2$ substrate

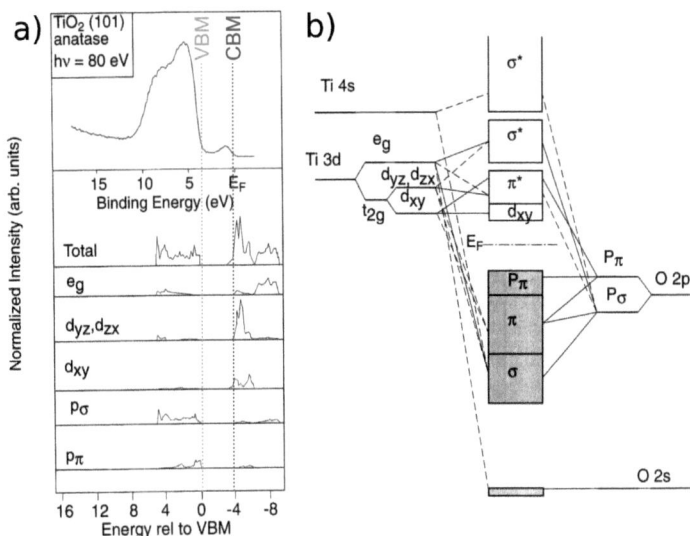

Figure 2.7: Calculations on the total and projected density of states (DOS) and the molecular orbital bonding structure of anatase (101) (modified from [130]). **a)** A valence band photoemission spectrum (top) compared to the projected density of states of the different TiO$_2$ molecular orbitals. The band gap has been corrected from 2 eV to 3.2 eV. **b)** An energy scheme of the molecular orbitals of TiO$_2$ showing strong (solid lines) and weak interactions (dashed lines) between atomic orbitals.

Ti 3d states. Hence the valence orbitals are mainly associated with O^{2-} ions, whereas the conduction band states are mainly associated with Ti^{4+} ions [40]. In the case of band gap excitation electron density is shifted from oxygen to titanium sites.

The interaction of the atomic orbitals of the anatase is displayed in the molecular orbital bonding diagram (Fig.2.7 b)). The upper valence band can be divided into three regions: the σ bonding in the lower part, a middle and a higher π bonding region reaching to the valence band maximum. Due to symmetrical reasons (distortion of the $[TiO_6]^{2-}$ octahedrons), not hybridized nonbonding molecular orbitals are located near the band gap, namely the O p_π orbital at the valence band maximum (VBM) and the Ti d_{xy} orbital at the conduction band minimum (CBM). The rest of the conduction band states is of antibonding character, mainly derived from Ti 3d and Ti 4s atomic orbitals.

Resonant Photoelectron Spectroscopy (ResPES; see Section 3.1) experiments have been conducted by Thomas et al. [131, 134] to investigate the hybridization of electronic states. The excitation energy was varied from 40–80 eV, which is around the Ti 3p $\longrightarrow$ Ti 3d resonance. Resonance of spectral features is attributed to a strong Ti–O hybridization [130, 135, 136]. By this technique the correctness of the molecular orbital bonding diagram (Fig.2.7 b)) has been confirmed. Furthermore the gap states could be attributed to Ti 3d states.

2.1.4 Defects

In the binary titanium-oxygen system a lot of phases with different stoichiometries exist [98]. Consequently a TiO_2 single crystal can be reduced easily e.g. upon annealing in vacuum. Annealing results in bulk defects, where electrons are located in an anion vacancy (color centers), which are capable to absorb visible light, thus changing the color of an initially transparent single crystal.

Unlike in single crystals, bulk defects become less important in nanoscopic anatase, but defects at the surface gain more significance. Therefore in the following only surface defect states are presented. Step edges on anatase are considered being the most common intrinsic defect yielding in fourfold coordinated 4c titanium. In addition it is expected that the 2c oxygen atoms can be removed easily upon annealing in UHV, which gives rise to point defects playing an important role in the surface science of TiO_2. In a STM study the point defects could be observed on the anatase (101) surface [123] and assigned to oxygen vacancies [122]. Accordingly this oxygen deficiencies were observed on rutile (110) single crystals [137].

It has been shown by Hengerer et al. [138], that upon sputtering the anatase phase at the (101) and (001) surface can be changed even to TiO because of preferential sputtering of oxygen atoms. In the case of the (101) surface the anatase phase could be restored completely at the surface without exposure to oxygen. This is due to point defect diffusion into the bulk either by oxygen vacancy or titanium interstitial diffusion. In contrast for the (001) surface exposure to oxygen was needed.

2.1 The TiO₂ substrate

To get more insights into the change of the electronic structure of anatase due to surface defect creation Photoelectron Spectroscopy (PES)(Section 3.1) is used as a powerful surface sensitive tool [22,139,140]. In a study of nearly defect free and defect rich nanocrystalline anatase samples formed by Ar ion sputtering (Fig.2.8), both X-ray Photoemission Spectroscopy (XPS) and Ultraviolet Photoemission Spectroscopy (UPS) was applied to investigate changes in oxidation state and valence band structure as well [22]. It has to be mentioned that UPS measurements with an excitation energy of $h\nu = 21.2$ eV is not as surface sensitive as Synchrotron Induced Photoelectron Spectroscopy used in this work. Hence the assignment of the unsputtered anatase being nearly defect free might be doubtful.

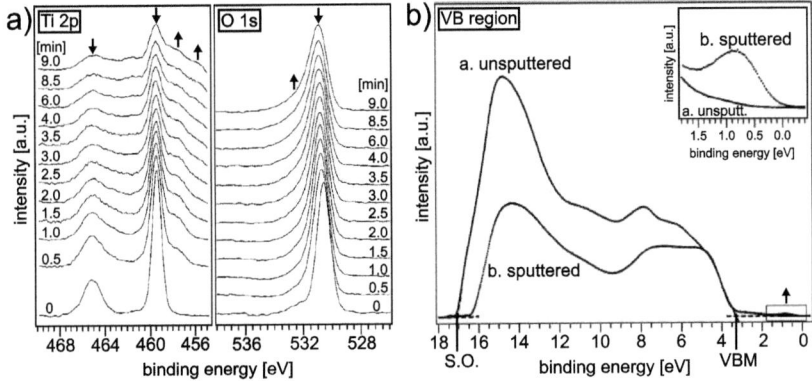

Figure 2.8: Photoemission spectra course of nanocrystalline anatase displaying sputtering effects [22]. **a)** X-ray Photoemission Spectroscopy(XPS) of the core levels Ti 2p and O 1s reveals dramatic changes with proceeding sputtering time. The arrows mark whether a feature gains intensity or decreases in intensity. **b)** By Ultraviolet Photoemission Spectroscopy (UPS) the creation of gap states formed by sputtering is shown (see inset).

The titanium Ti 2p core level spectra (Fig.2.8 a)) for the untreated TiO₂ show two emission lines of Ti $2p_{3/2}$ at 459.4 eV and Ti $2p_{1/2}$ at 465.1 eV showing a spin orbit splitting of 5.7 eV. This emissions can be unambiguously assigned to the Ti^{4+} oxidation state. With increased sputtering additional peaks appear at lower binding energy (Ti $2p_{3/2}$ = 457.5 eV and Ti $2p_{1/2}$ 463.2 eV) being attributed to Ti^{3+} states. At longer sputter exposure time Ti^{2+} states can be observed as well at even lower binding energies (Ti $2p_{3/2}$ = 455.6 eV and Ti $2p_{1/2}$ 461.3 eV). Generally the reduced Ti 2p components were observed throughout literature [98, 102,131,139,141–143]. The oxygen O 1s emission line consists of three components: One main emission line being assigned to bulk TiO_2 at 530.6 eV, and two smaller contributions being located as a wide shoulder at higher binding energies. A component at 531.5 ± 0.5 eV is attributed to bridging oxygen (2c-O) and at 533 eV due to OH adsorption. The relative intensity of the O 1s emission decreases, which indicates clearly that upon sputtering oxygen is removed preferentially. Figure 2.8 b) shows the formation of band gap states (BE ~ 0.8 eV) in UPS. Since the formation of the band gap states comes along with an intensity decrease of the O 1s emission line (Fig.2.8 a) right), one can correlate these band gap states to oxygen vacancies.

The existence of these oxygen vacancies and their correlation to occupied Ti 3d states has been proved in several publications [98, 131, 134, 140, 142, 143]. The creation of oxygen vacancies can be described by

$$TiO_2 \longrightarrow TiO_{2-x} + xV_O + \frac{1}{2}xO_2 \qquad (2.1)$$

where V_O represents an oxygen vacancy and x the fraction of oxygen vacancies in the material. The amount of oxygen vacancies can be controlled by the exposure to oxygen [131].

Unlike these band gap spectra, published results by Westermark et al. [21] and the results in this work (Page 90 Fig.6.10) show additional band gap states just below the Fermi level, referred to as $Ti3d < E_F$ states. Their origin is not yet clarified. These states have not been observed on single crystalline anatase [131]. They might arise from the Ti d_{xy} orbitals, which are not hybridized.

The role of defects for the performance of the DSSC is still under debate. It is widely accepted, that defect states (surface states) within the gap act as recombination sites [48, 76, 78, 144, 145], thus playing a detrimental role for the DSSC. It it assumed, that electrons can recombine with the oxidized I_3^- species of the electrolyte . But some publications state that photoexcitation by UV light illumination produces surface states, which are beneficial for electron transport [17, 80], if the density of these surface states is sufficiently large. The conduction band is shifted to higher binding energies, which increases the driving force for electron injection from the dye into the TiO_2. Depending on the sensitizing dye, the increase in efficiency of the DSSC ranges from two to 45 orders of magnitude [79].

2.2 The N3 Ru dye

For efficiently sensitizing wide bandgap semiconductors like anatase, it is necessary to employ a dye extending the incident photon to current efficiency (ICPE) into the visible and near infrared solar spectrum (Fig.2.9). The IPCE is the number of electrons generated in the external circuit divided by the number of incident photons depending on the excitation wavelength. Another very important feature of the dye is a high stability in the oxidized state (see Section 1.3.2). In particular ruthenium (II) complexes have been applied largely for dye cells. This is due to the following reasons [146]: At first, the octahedral geometry of Ru complexes allows to attach ligands in a controlled way for tuning the dye towards desired (photo)physical and chemical properties in a predictable way. Secondly, the oxidation states from I to IV are stable within the metal complex. Besides that the dye is suitable for anchoring on the TiO_2 surface (Section 2.2.2) and ensures an ultrafast electron injection into the semiconductor (Section 1.3.1).

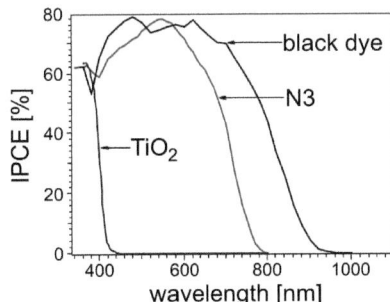

Figure 2.9: The IPCE as a function of the incident photon wavelength (from [10]). For comparison the unsensitized TiO_2 is also shown.

The most widely used sensitizer in the DSSC is the dye $RuL'_2(NCS)_2$ (Ru^{II}(2,2'-bipyridine-4,4'-dicarboxylate)$_2$(NCS)$_2$), being referred to as N3 in the following. The Ru dye N3 is still among the stable dyes leading to highest efficiencies in the DSSC [18]. The N3 dye is only outperformed in efficiency by the black dye (Ru^{II}(4,4',4"-tricarboxy-2,2':6',2"-terpyridine)(NCS)$_3$), which pushes the absorption onset of light 100 nm further into the infrared region. Research has been extended to dye complexes with different central metal atoms [147], but the Ru complexes are still preferred, because of the ease of tunability of these complexes. Also different ligands have been investigated [148].

2.2.1 Molecular and electronic structure

According to crystal field theory in an octahedral metal complex the d-orbitals of the Ru central atom split into degenerate t_{2g} (d_{xy}, d_{xz}, d_{yz}) and e_g ($d_{x^2-y^2}$, d_{z^2}) sets of orbitals. In the Ru N3 dye the Ru atom is coordinated by two bivalent bipyridine and two monovalent thiocyanate (SCN) ligands (Fig.2.10 a)). The coordination octahedron is slightly distorted [149].

On the N3 dye molecule the HOMO and LUMO orbitals are spatially separated from each

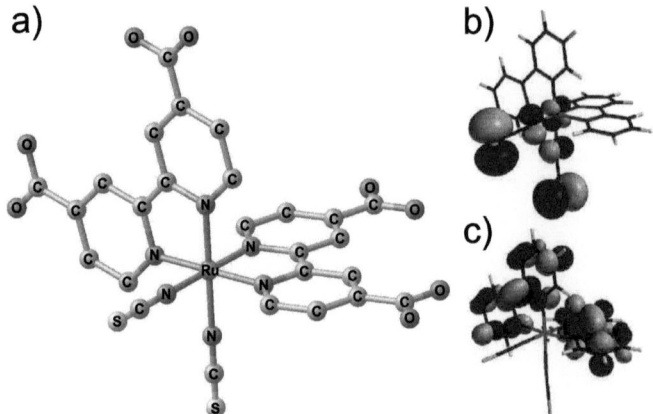

Figure 2.10: The RuL'$_2$(NCS)$_2$ molecule, also known as the N3 dye. **a)** Model of the molecular structure showing the different atoms within the molecule. **b)** The calculated HOMO and **c)** LUMO orbital surfaces [9, 150].

other [9, 150, 151]. One can see that the HOMO of the dye is mainly located on the thiocyanate group, whereas the LUMO is spread over the π-electron system of the bipyridine ligands.

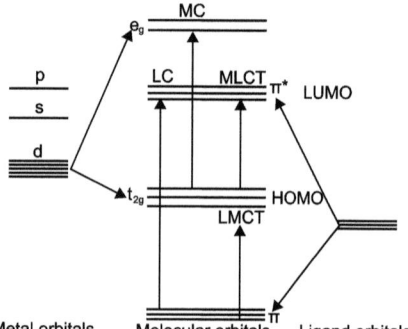

Figure 2.11: Electronic scheme of the molecular orbitals created by metal and ligand atomic orbitals. In this scheme possible transitions are indicated by arrows (adapted from [146]).

A deeper understanding of the photophysics of the dye can be obtained by an electronic energy diagram of the dye (Fig.2.11). Three excited states exist in these metal organic complexes leading to four possible charge transitions: The first is metal centered (MC), which is due to an excitation of an electron from t_{2g} to e_g states. The second is ligand centered (LC) because of a $\pi \rightarrow \pi^*$ transition. The third one is originating from a metal to ligand charge transfer (MLCT). Overall the two latter transitions are most relevant for device operation, since they contribute to the vectorial charge transfer. The MLCT transition stands for the optical excitation of an electron from the HOMO level to the LUMO by incident light. In other words, the light absorption is described by this type of charge transition within the molecule. The fourth charge transition is due to a ligand to metal charge transfer (LMCT). The LMCT is not discussed for the RuL'$_2$(NCS)$_2$. The t_{2g} states are allocated as the HOMO of the dye and therefore occupied. Thus electrons cannot be excited into the t_{2g} states.

Detailed density functional theory (DFT) calculations of the respective electronic ground and excited states of the N3 dye molecule have been accomplished [152–154]. Moreover time dependent density functional theory calculations on the electronic states and their dependence on protonation were conducted [155, 156]. In Ref. [156] the line up with the band edges of a model TiO$_2$ nanoparticle consisting of a Ti$_{38}$O$_{76}$ cluster (from Ref. [157]) exposing anatase (101) surfaces is calculated. By decreasing protonation of the dye the HOMO and LUMO energy move towards the vacuum level, while the distance between both levels increases. This increase of the gap between HOMO and LUMO leads to a blue shift observed in optical absorption spectra [158]. For a protonated dye upon adsorption the anchoring carboxyclic groups deprotonate and donate their proton to the semiconductor charging it positively. This might lead to a downward shift of the Fermi level (*Nernst shift*; see Section 1.2) [33–35] and enhances electron injection into the semiconductor implying higher photocurrents. On the other hand the gap between the redox couple and the Fermi level decreases, which causes a lower open circuit potential. Thus an optimal degree of protonation of the dye is needed for maximizing the product of short circuit current and open circuit potential.

To investigate the energy alignment of the HOMO level cyclic voltammetry as well as photoelectron spectroscopy have been performed by several groups (see Section 1.1).

2.2.2 Adsorption geometry on anatase

Due to the distinct geometry of the dye molecule its orientation relative to the TiO$_2$ semiconductor is of prime importance for the performance of the Dye Sensitized Solar Cell. Since the anatase (101) is the thermodynamically favorable surface as already mentioned above (Section 2.1.1), it will be exclusively in the focus of the following geometrical adsorption considerations. Taking the distances of the possible adsorption sites into account on the part of the dye molecule and the anatase (101) surface (Fig.2.12 a)), one comes up with a variety of possible adsorption geometries. The N3 dye complex anchors preferably with the carboxyl groups on unsaturated 5c-Ti sites (Fig.2.12b) – d)), but an interaction between TiO$_2$ substrate and thiocyanate group (Fig.2.12 e) and f)) is suggested in the literature as well [159].

As the initial step of the dye anchoring to the TiO$_2$ surface in Ref. [149] the coupling via only one carboxyl group is proposed. In subsequential anchoring steps the dye can adsorb with any of the others remaining uncoupled carboxyl groups, which can end up to any of the shown adsorption geometries (Fig.2.12b) – f)). Experimental support for the bonding of two carboxyl groups per dye molecule has been provided by Infrared Analysis (IR) [161]. The coupling via two carboxyl groups of the same bipyridine ligand (Fig.2.12 b)) allows the molecule to tilt sideways, which may lead to the configuration shown in Fig.2.12 e). The

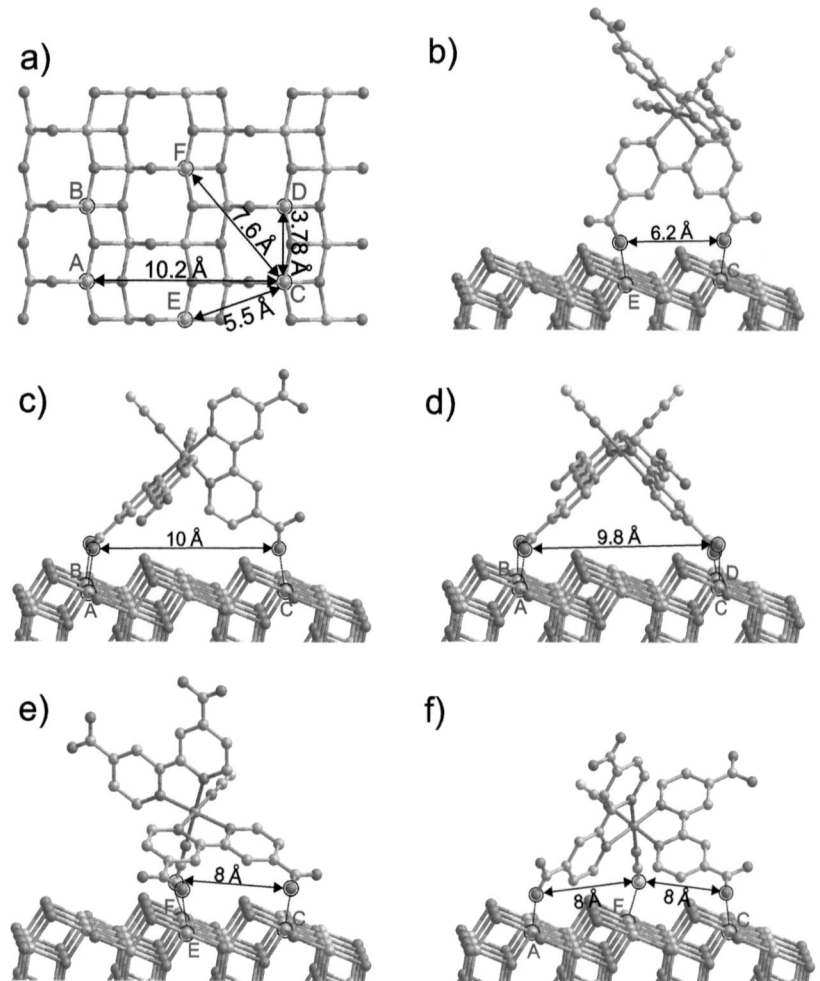

Figure 2.12: Five possible adsorption geometries of RuL'$_2$(NCS)$_2$ dye on the anatase (101) surface (drawn accordingly to [149, 160]). **a)** Topview onto the (101) anatase surface with the possible adsorption sites for the RuL'$_2$(NCS)$_2$ dye. For a better view only the first layer of atoms is shown. **b)** Sideview onto the (101) anatase surface. The carboxyl groups of only one bipyridine ligand are coupled to the surface with two oxygen atoms. **c)** The N3 dye is bound to the substrate by three oxygen atoms. **d)** Here the dye is bound with four oxygen atoms to the substrate. **e)** This adsorption geometry originates from Fig. 2.12b) and involves the interaction with the thiocyanate (SCN) groups. The molecule is tilted in a way that the SCN group can interact with adsorption site F. **f)** Two oxygens of the carboxyl group of different bipyridine ligands and one SCN group anchor to the surface.

2.2 The N3 Ru dye

mismatch between the distance of the anchoring oxygen atoms of the dye (O–O = 6.2 Å) compared to the best suitable Ti–Ti distance (5.5 Å) is much larger than the mismatch for the adsorption geometries shown in Fig.2.12 c) and d), which may lead to different strains induced by relaxation upon these mismatches [162]. In an experimental study K-edge X-Ray Absorption Fine Structure (XAFS) data show a higher degree of disorder in dye sensitized nanocrystalline anatase films than in blank ones [163].

Shklover et al. [149] suggest that the large mismatch makes the configuration shown in Fig.2.12 b) improbable. In Fig.2.12 c) and d) the molecule is inhibited to tilt due to a stronger coupling to the substrate by more oxygen atoms, which make an interaction between the thiocyanate group and substrate unlikely. Shklover et al. [149] state, that the adsorption geometry shown in Fig.2.12 d) is most favorable by thermodynamic reasons, but a direct evidence is missing.

The N3 dye molecule has a distinct geometry, which guarantees, that the dye operates like an electronic membrane. Due to the spatial separation of the HOMO (Fig.2.2 b)) and LUMO (Fig.2.2 c)) an effective vectorial charge transfer of electrons and holes takes place [1] for the dye adsorbed like in Fig.2.12. The vectorial charge transfer can be explained in terms of Fermi's Golden Rule (Section 1.3.1, Equation 1.3.1) in a way, that the injection rate k_{inj} is governed by the matrix element M_{fi}. It is large, if the initial state, (the LUMO level of the dye), and the final state (the TiO_2 conduction band states) have an overlap. Therefore a strong coupling of the carboxyl groups to the TiO_2 substrate is favorable. In contrast the interaction between the thiocyanate groups and the substrate reduces the vectorial charge transfer of the dye and interferes with its function as an electronic membrane. In this case the matrix element M_{fi} is small. A photoelectron spectroscopic study of dry dye adsorbed on nanocrystalline anatase [159] suggests a fraction of about 60% of the N3 dye molecules on the TiO_2 surface showing an interaction of the thiocyanate group with the substrate.

2.2.3 Anchoring modes onto anatase

In the following the anchoring mechanism of the carboxyl group onto the TiO_2 substrate will be examined in detail (Fig.2.13). Assuming a deprotonation of the anchoring groups from IR analysis [164, 165], three binding modes of the carboxyl group onto unsaturated 5c-Ti sites are possible leading to dissociative adsorption. So far experimental evidence for the deprotonation of formic acid[1] on defined surface orientations is present only on rutile (110) [166, 167]. But IR data of carboxylated ruthenium dyes on nanocrystalline TiO_2 indicate, that deprotonation occurs for anatase (101) surfaces as well [164, 165].

In all adsorption geometries shown in Fig.2.12, all three binding modes (Fig.2.13) are possible in principle, because the carboxyl groups are able to rotate around the C–C axis, which attaches them to the bipyridine ligand. But there is no experimental evidence given for

---

[1] The most simple molecules containing one carboxyl group is the formic acid (HCOOH). It can be used as a probe for the reaction of carboxyl groups on TiO_2.

Figure 2.13: The different coupling modes of the carboxyl group to the TiO$_2$ substrate [165, 168]. For simplicity, the unbound electron pairs at the oxygen atoms are omitted.

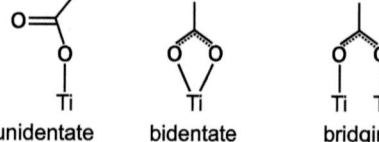

the creation of an unidentate linkage to the TiO$_2$ substrate [165]. Of the two remaining anchoring modes the bridging one has been found to be the most stable configuration [164].

In an earlier theoretical study besides the dissociative adsorption also the molecular adsorption of dye molecules without deprotonation of the carboxyl groups was considered [169]. But lacking experimental evidence it is not assumed that the carboxyl groups undergo adsorption without deprotonation.

A beneficial side effect of the coupling of the dye onto the TiO$_2$ surface has been found. In solution the N3 dye molecule suffers from degradation of the thiocyanate ligand under irradiation, i.e. sulfur loss. But upon anchoring to the substrate, this degradation mechanism is suppressed due to the rapid transfer of the excited electron into the semiconductor, which deactivates the excited state of the molecule [170]. The electron injection rate constant is estimated being 10^9 times larger than the sulfur loss reaction.

2.3 The electrolyte

The electrolyte in the Dye Sensitized Solar Cell works equivalent to a hole conductor, which consists of the I^-/I_3^- redox couple mainly used in the dye cell, the cation of the redox couple, and the solvent. In the literature this redox couple is often termed as a (redox) mediator, i.e. providing a charge flow between working and counter electrode within the device.

The electrolyte has to match a couple of requirements to provide the needed functions [171]. As a basic prerequisite the electronic alignment of the redox couple relative to the dye (Section 1.1) must allow a rereduction of the dye by means of electron donation from the reduced I^- (Section 1.3.2). The more negative the redox potential of the mediator is situated, the larger is the thermodynamic driving force of this rereduction process. In addition a lot more conditions have to be fulfilled: The redox couple ions have to be soluble in the used solvent. Furthermore a high diffusion coefficient for a fast ion transport is demanded. The redox system must possess a high reversibility of involved charge transfer reactions. On the other hand a high stability of the reduced (I^-) and oxidized (I_3^-) state within the solvent is also very important. To avoid loss of incident light, the electrolyte should not absorb light in the visible region. Last but not least in order to prevent undesired side reactions a high resistivity of the electrolyte against ambient conditions is necessary.

Drying-up of the electrolyte is the main issue concerning long term stability of the cell. Therefore many efforts have been attempted to get rid of the liquid electrolyte employing a solid hole conductor (Section 1.4) or ionic liquids without the volatile solvent [133, 172–177]. A very interesting approach has been reported very recently employing a liquid crystal electrolyte [178]. Despite the high viscosity of the liquid crystalline electrolyte, the I^-/I_3^- redox couple moves within ion conducting layers of the smectic phase formed by the cations [179,180]. However, DSSCs employing liquid electrolytes yield overall still higher efficiencies than highly viscous electrolytes or solid hole conductors.

2.3.1 Solvation of the electrolyte ions

Solvation processes are governed by intermolecular interactions, which are different for polar and unpolar solvents. Hydrogen bonding, ion–dipole and dipole–dipole interactions occur only in polar solvents, whereas van der Waals forces and ion–ion interaction are present in unpolar solvents. For calculating the Gibbs free energy of solvation, the first theoretical concept was the *Born model* [181]

$$\Delta G = -\frac{Z^2 e^2}{8\pi\epsilon_0 R}\left(1-\frac{1}{\epsilon}\right) \qquad (2.2)$$

with Z, e, R, ϵ_0 and ϵ being the charge of the ion, the unit charge, the radius of the ion, the dielectric constant of vacuum and of the solvent respectively. In Table 2.2 calculated values of the solvation enthalpy ΔG for the anions and cations are given. The dielectric constants are for acetonitrile $\epsilon = 38$ and for benzene $\epsilon = 2.3$ [182].

	I^-	I_3^-	Li^+	$PMII^+$
ionic radius [pm]	206	580	73	570
$\Delta G_{acetonitrile}$ [J mol^{-1}]	-327.6	-116.4	-924.5	-118.4
$\Delta G_{benzene}$ [J mol^{-1}]	-190.2	-67.5	-536.6	-68.7

Table 2.2: The solvation enthalpies for the redox couple I^-/I_3^- and the cation Li^+ in J mol^{-1}, according to the Born model (Eqn.2.2). The values for the ionic radii are taken from [183]. The ionic radius of the $PMII^+$ (1-propyl-3-methylimidazolium iodide) cation has been estimated by its molecule structure (see Figure 2.17).

As one can see, the Gibbs energy of solvating of the smaller Li^+ cation is much higher than for the I_3^- and for the I^- anion. Due to the much higher dielectric constant of acetonitrile compared to benzene the first is a better solvent for ions. The Born model treats the solvent as an uniform continuum embedding the solvated ion in a spherical cavity. Hence it neglects the structure of the solvation shell due to immobilized solvent molecules. This shortcoming has been overcome by accounting for the interaction within the first solvation shell by Marcus [184].

But as it has been shown in recent Photoelectron Spectroscopy experiments conducted by Weber et al., the Born model reproduces reasonably well absolute binding energy values of iodide anions and alkali metal cations dissolved in water [185, 186]. In an earlier work accomplished by Mayer the same behavior was found for bromide anions and sodium cations in water [187]. In both studies the effect of solvation has been investigated by showing binding energy shifts of opposite direction for the dissolved cations and anions relative to gas phase spectra. For cations the gas–liquid shift goes to lower binding energies, whereas the anions show a shift towards higher binding energies. Because of the opposite direction of the solvent dipoles (Fig.2.14), the binding energies are shifted in different directions, which can be observed by Photoelectron Spectroscopy.

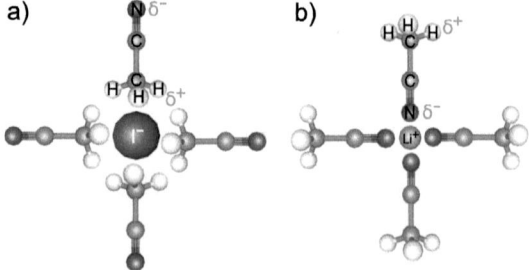

Figure 2.14: Scheme of the solvation in acetonitrile of an **a)** iodide anion **b)** lithium cation. Because the anion is much larger than the cation, the solvation cage is larger as well (not drawn to scale).

Using photoemission experiments combined with molecular dynamics calculations, Markovich et al. [188] investigated the coordination of acetonitrile solvent molecules around I^- ions. In this study the stabilization energy due to solvation of I^- ions by acetonitrile was found

2.3 The electrolyte

to increase with the number of surrounding solvent molecules. The calculations and experiments reveal a strong initial change of the stabilization energy up to 12 molecules. Thus the authors conclude, that the first solvation layer consists of preferably 12 acetonitrile molecules. Afterwards the stabilization energy further increases with the number of solvating acetonitrile molecules, but now much slower as the initial change.

2.3.2 Electronic structure

Caused by electrostatic interactions of the ion with the solvent molecules, a solvation shell of the solvent molecules around the ion is formed. Depending on the charge on the solvated ion, the dipoles of the solvent molecules are arranged in a solvation cage. The energy levels for the oxidized and reduced species of the I^-/I_3^- redox couple shift by a reorganization energy λ_{ox} and λ_{red} relative to the redox potential E^0_{redox} respectively [189], as is depicted in Fig.2.15 for a reversible one electron transfer redox couple. The reorganization energy is due to a *Franck-Condon split*, because the rearrangement of the molecules lags behind the fast charge transfer reaction, which is described in detail by the well established *Marcus theory*. One has to consider, that the corresponding energy states of the oxidized and reduced state belong to different species and are sterically apart from each other. Thus no direct charge transition between the two energy state distributions is possible, as one could misinterpret from Fig.2.15. Usually λ is in the range of 0.5–2 eV [190], which is depending on the strength of interaction between solvent and ions. The energy levels are broadened because of thermal fluctuations of the solvation shell, which is assumed to result in a Gaussian type of distribution. As a consequence occupied and empty states, and their density of states (DOS) as well, correspond to reduced and oxidized species of the redox system [190]. It should be noted that for the I^-/I_3^- redox couple the energy distribution of energy states considerably deviate from this simplified picture, because different chemical species and related energy states are involved.

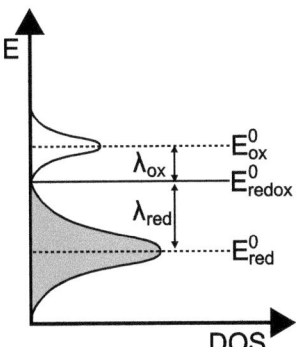

Figure 2.15: The occupied reduced (E^0_{red}) and unoccupied oxidized (E^0_{ox}) states of an ion in solution (after [187]). The standard redox potential is given by E^0_{redox} and the reorganization energy by λ_{ox} and λ_{red} respectively.

According to Figure 2.15, the redox potential of the electrolyte E_{redox} is depending on the concentration of the reduced and oxidized species. Quantitatively this is described by the *Nernst equation*:

$$E_{redox} = E^0_{redox} + \frac{RT}{nF} \ln \frac{c_{ox}}{c_{red}} \qquad (2.3)$$

E^0_{redox} is the standard potential, R the universal gas constant, T the temperature, n the number of transferred electrons, F the Faraday constant and c_{ox} and c_{red} the concentration of the oxidized and reduced redox species. Respective to this equation, the redox potential E_{redox} shifts by 59 meV per decade of the concentration ratio of $\frac{c_{ox}}{c_{red}}$. Thus depending on this concentration ratio the driving force for the rereduction of the dye varies.

It is expected that the I^-/I_3^- redox couple has a more negative potential in acetonitrile due to solvation effects compared to aqueous solution (+0.536 V vs. NHE). Quite reasonable determined values were published by two groups ranging from +0.256 V [191] to $\sim$ +0.27 V vs. NHE [192]. For the alignment of the electronic levels of all components contained in the DSSC (Figure 1.1), a value of $\sim$ +0.4 V vs. NHE has been taken.

2.3.3 The I^-/I_3^- redox couple

The standard mediator used for the DSSC is the I^-/I_3^- redox couple. Of course some groups searched for alternatives like metal organic cobalt complex redox couples [193–196], thiocyanate, bromide or ferrocene [171]. But since O'Regan and Grätzel introduced the nanocrystalline DSSC [6], no other redox couple has exceeded the I^-/I_3^- redox couple in efficiency [171].

The function of the redox couple in the DSSC is to rereduce the oxidized dye, which already has been described in Section 1.3.2. Transient absorption experiments performed by Clifford et al. [197] indicate, that an intermediate [I^- dye$^+$] complex is formed in order to donate an electron from the solvated I^- ion to the thiocyanate group (location of the HOMO).

Iodine (I_2) has only a small solubility in polar solvents (0.3 g kg^{-1}), but in solutions containing a dissolved iodine salt (including I^- ions) it can be dissolved easily. In solution the equilibrium reaction

$$I^- + I_2 \rightleftharpoons I_3^- \qquad (2.4)$$

occurs. Although the existence of this I_3^- complex was discovered in the year 1819, another 150 years passed by to understand the binding within this complex, which can be referred to in usual textbooks [183, 198]. Although the existence of polyiodide anionic species like I_5^-, I_7^-, I_9^-,... has been a doctrine over decades, the formation of such complex has been scrutinized recently due to missing experimental evidence [199].

2.3.4 The cation

The cation also has a big influence on cell performance. It screens electrons inside the nanoparticles (Section 1.3.3.1), influences the electron injection yield [37, 200], the electron diffusion coefficient [51] and the dye rereduction rate (Section 1.3.2) [201]. All these aspects contribute to the current density, which rises with decreasing cation size [171, 200], as being depicted in Fig.2.16 a).

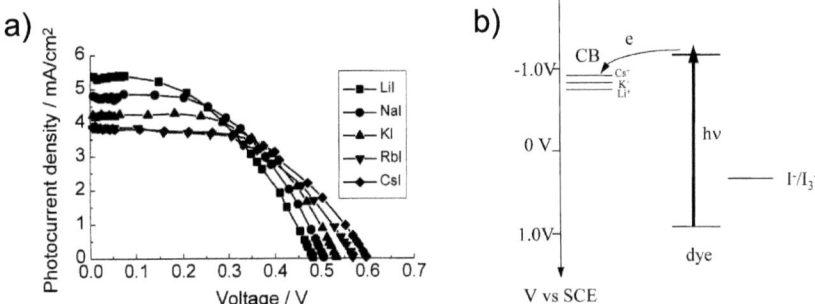

Figure 2.16: The cation size plays an important role for cell performance (taken from [200]). **a)** The current voltage characteristic depend largely on the cation size. The photocurrent density is higher for smaller cations, whereas the open circuit voltage increases for larger cations. **b)** The conduction band edge (CBE) potential is lowered relative to the electronic dye levels.

By adsorption of cations on the TiO_2 surface the driving force for charge injection is increased since the conduction band edge (CBE) potential shifts downwards relative to the LUMO of the dye (Fig.2.16 b)), which has been observed in several studies [200, 202–205]. In addition, the smaller the cation the better it can penetrate the dye layer on the surface, transferring the double layer potential drop closer to the TiO_2 surface (see Section 1.2, Figure 1.3). On the other hand the open circuit voltage is decreasing with decreasing cation size [31, 85, 200] because of the same reason.

The electron diffusion coefficient is lowered by larger cations, because the cation has to codiffuse with the I^- anions into the TiO_2 in order to maintain charge neutrality.

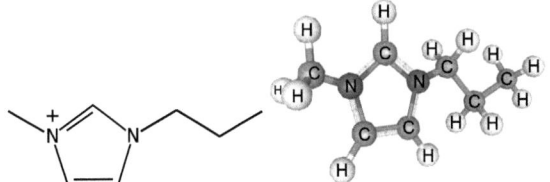

Figure 2.17: left: The molecule structure of 1-propyl-3-methylimidazolium. right: 3D view of the molcule with attached hydrogen atoms.

However, using molten salts (ionic liquids), which have much larger cations compared to alkali cations, results in a lower photocurrent, but a higher photovoltage and fill factor. Besides Li, the cation from the molten salt 1-propyl-3-methylimidazolium iodide (PMII) has been used in this present work (Fig.2.17). Overall the efficiency of solar cells can be significantly increased [206]. Even without solvent reasonable conversion efficiencies with ionic liquids up to 7.4% have been reached [133]. Overall ionic liquids seem to be a promising alternative.

2.3.5 The solvent

The solvent is ideally aprotic to avoid photochemical degradation reactions in particular with the dye molecule. Especially the thiocyanate ligand is sensitive to the ambient and UV-radiation, e.g. due to an exchange reaction of this ligand with water molecules [207]. Since the thiocyanate ligand is responsible for the electron uptake for rereduction, this degradation corrupts cell performance. In addition water as a solvent for the redox couple causes an undesired blue shift of the dye, which narrows the spectrum of absorbed incident light [208]. A desired property of the solvent is a low viscosity, which facilitates a high ion mobility [209]. Overall the solvent should provide an inert environment for the redox couple to stabilize the electrolyte in harsh conditions like high thermal and UV input due to sunlight irradiation.

2.3.5.1 Acetonitrile

Acetonitrile (CH_3CN) is a colorless aprotic solvent with high polarity (dielectric constant $\epsilon = 38$ at 25 °C) [182]. It is a volatile solvent with a low viscosity, thus it is a suitbale solvent for the I^-/I_3^- redox couple ensuring a high ion mobility. The high polarity (a high ϵ) is responsible for a large solvation enthalpy, as shown in Table 2.2). Since the solvation behavior for the redox couple in acetonitrile has been described already, we will now focus on solvent–substrate interactions.

Adsorption behavior Acetonitrile is used as a suitable probe for testing Brønsted or Lewis sites on various materials analyzing its CN stretching frequency ($\nu(CN)$) [210–212]. It is well known, that the frequency (2254 cm^{-1}) shifts to higher values upon interaction of the nitrogen lone pair with electron-accepting adsorption sites on the substrate. A suggested adsorption mechanism based on infrared spectroscopy of acetonitrile adsorption onto TiO_2 (nanocrystalline anatase) is shown in Fig.2.18.

The $\nu(CN)$ at 2274 cm^{-1} is related to the Brønsted acid site bonding (proton donation from the adsorption site), whereas the stronger shift ($\nu(CN) = 2318$ cm^{-1}) is attributed to a Lewis acid site bonding. Thus the interaction between acetonitrile and TiO_2 is stronger at the Lewis site, indicating a stronger bonding. In this study also the adsorption behavior for defect rich (Ti^{3+} rich) and nearly defect free samples with adsorbed acetonitrile was investigated. The adsorption on Ti sites with different oxidation state should lead to different CN stretching frequencies, but for the two differently treated samples the infrared spectra at $\nu(CN)$ look the same. Therefore the authors conclude, that no adsorption on

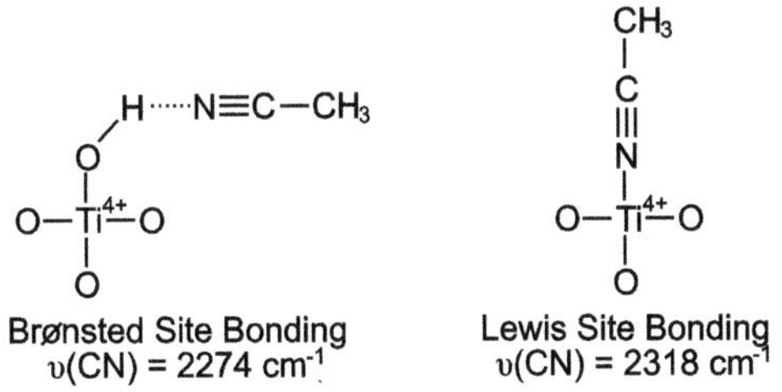

2.3 The electrolyte

Figure 2.18: The adsorption mechanism of acetonitrile onto TiO_2 (from [213]). **left:** A smaller shift of the CN stretching frequency is attributed to Brønsted acid site bonding. **right:** The Lewis acid site bonding is deduced from a higher frequency shift.

Brønsted Site Bonding $\upsilon(CN) = 2274\ cm^{-1}$

Lewis Site Bonding $\upsilon(CN) = 2318\ cm^{-1}$

Ti^{3+} sites occur, because these sites are possibly not accessible for acetonitrile due to sterical reasons. This conclusion is in contradiction to the experimental findings in this work, which are shown in Section 7.

Hydrolysis and photodissociation Water contamination within acetonitrile is harmful for the ruthenium dye complex, because an exchange reaction of the thiocyanate group and water occurs [207]. But even for acetonitrile itself addition of water can cause decomposition. Measurements using infrared spectroscopy indicate that on oxides like zirconia (ZrO_2) and TiO_2 water causes a hydrolysis reaction of acetonitrile, being bound on the oxide surface [214, 215].

It is widely accepted, that without water photocatalytic degradation of acetonitrile on TiO_2 occurs only in the presence of free oxygen gas [216–219]. But Chuang et al. reported a decomposition reaction due to hydroxyl groups on the surface of TiO_2 without having free oxygen upon strong irradiation with UV light [220]. In contrast to that, Zhuang et al. state that hydroxyl or oxygen on top of the TiO_2 surface is not sufficient for a photodissociation of acetonitrile [213].

2.3.5.2 Benzene

Since the solvation behavior is dependent on the polarity of the solvent (see page 45, Equation 2.2), benzene was chosen as an additional solvent for comparison. In contrast to acetonitrile it is nonpolar and has a very low dielectric constant ($\epsilon = 2.3$), thus the solvation behavior of iodide is completely different. The solvation enthalpy ΔG is much smaller (see. Table 2.2). No solvation shell with oriented solvent dipoles around the ions is built up due to missing molecular dipoles. Benzene (C_6H_6) is, like acetonitrile, a colorless, volatile and aprotic solvent having a low viscosity.

Adsorption behavior Like in the case of acetonitrile adsorption infrared measurements have shown a stronger interaction of benzene on the Lewis than on the Brønsted site [221, 222]. Since benzene molecules interact with their π electron system with the substrate surface, the benzene rings adsorb lying flat on the surface, which has been shown by

Near Edge X-ray Absorption Fine Structure (NEXAFS) experiments [223] and theoretical calculations [224, 225].

Photodissociation Photodissociation of benzene has been reported upon X-ray irradiation in the presence of oxygen forming carbon dioxide [226]. It is also indicated by XPS measurements, that X-ray irradiation at room temperature causes polymerization of benzene on TiO_2, whereas at lower temperatures (120 K) it is stable [227].

Part II

Experimental

3 Experimental characterization methods

In this chapter the different characterization techniques used in this work are introduced. The emphasis is on the Photoelectron Spectroscopy (PES), which was used throughout the work as the main characterization method. For characterization of the TiO₂ substrate, different methods, like Atomic Force Microscopy (AFM), Scanning Electron Microscopy (SEM), Grazing Incidence X-ray Diffraction (GIXRD) and Raman Spectroscopy have been applied.

3.1 Photoelectron Spectroscopy (PES)

Photoelectron Spectroscopy [228, 229] was developed by Kai Siegbahn since 1960, also known as Electron Spectroscopy for Chemical Analysis (ESCA). This method is based on the outer photoelectric effect, which was discovered by Heinrich Hertz in 1886 [230], described in detail by Wilhelm Hallwachs in 1887 [231] and theoretically explained by Albert Einstein in 1905 [232]. Photoelectron Spectroscopy is one of the most important techniques in surface science. It is a powerful tool for characterizing the electronic structure, chemical composition and bonding interactions of materials with highest surface sensitivity. Photoelectron Spectroscopy is described in detail in the literature [233–236].

Working principle Upon irradiation with monochromatized electromagnetic radiation of energy $h\nu$, the atoms of the sample absorbs these incident photons. Electrons are excited from initial (occupied) states into final (unoccupied) electronic states. The incident light must have a higher energy than the work function

$$\Phi^S = E_{vac} - E_F \qquad (3.1)$$

of the investigated material in order to emit electrons out of the material with a kinetic energy E_{kin}^S, where E_{vac} is the vacuum level and E_F the Fermi level. This emitted electrons can be analyzed by a hemispherical analyzer. By an electrostatic lens system the counter voltage is varied to change the kinetic energy of electrons reaching the analyzer while measuring a spectrum. The electrons are brought to a defined kinetic energy E_{pass} passing through the analyzer by the hemispherical analyzer. Hence an energy dispersive analysis is possible (Fig.3.1 a)). The energy relations for Photoelectron Spectroscopy are displayed in Fig.3.1 b). The kinetic energy of an emitted core level electron is given by the excitation energy minus the core level binding energy $E_{B,CL}$ referenced to the Fermi level and the work function of the sample ϕ^S:

$$E^S_{kin,CL} = h\nu - E_{B,CL} - \phi^S \tag{3.2}$$

The Fermi levels of the sample and the analyzer are aligned to each other building up a contact potential $\Delta\phi$ due to different work functions between sample ϕ^S and analyzer ϕ^A, which accelerates or slows down the photoemitted electrons. Hence the kinetic energy of electrons leaving the sample $E^S_{kin,CL}$ is different from the kinetic energy in the analyzer $E^A_{kin,CL}$. The kinetic energy in the analyzer is

$$\begin{align} E^A_{kin,CL} &= h\nu - E_{B,CL} - \phi^S - (\phi^A - \phi^S) \tag{3.3} \\ &= h\nu - E_{B,CL} - \phi^A \tag{3.4} \end{align}$$

thus being independent of the work function of the sample. The binding energy of electrons at the Fermi level E_F is zero by definition, which makes an energy calibration possible. Knowing the excitation energy $h\nu$ the work function of the sample $\phi^S = h\nu - E_{SE}$ can be determined by the energetic position of the secondary edge E_{SE}, at which the kinetic energy of electrons is zero.

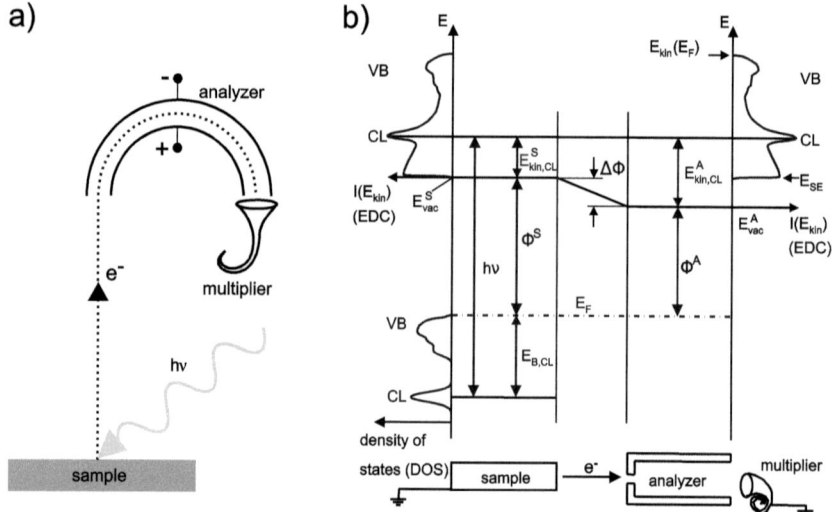

Figure 3.1: a) The working principle of Photoelectron Spectroscopy (PES). **b)** Energy scheme of the involved energy levels in Photoelectron Spectroscopy (from [237]).

In Photoelectron Spectroscopy the number of photoemitted electrons is measured in dependence on their kinetic energy, which results in an energy dispersion curve (EDC). This EDC resembles the density of states (DOS) in the material. In addition, a background of inelastically scattered electrons (secondary electrons) is superimposed in the EDC. This

3.1 Photoelectron Spectroscopy (PES)

background increases in intensity towards lower kinetic energy and breaks down to zero at zero kinetic energy (at the secondary electron edge). The line width of the emission lines is determined by the natural line width of a transition due to the finite lifetime of the photohole ΔE_{LT}, the line width of the radiation source $\Delta E_{h\nu}$ and the resolution of the analyzer ΔE_A. The first natural line width causes a Lorentzian broadening, whereas the two last contributions are of Gaussian type. The overall resolution Δ_{FWHM} is given by

$$\Delta E_{FWHM} = \sqrt{\Delta E_{LT}^2 + \Delta E_{h\nu}^2 + \Delta E_A^2} \qquad (3.5)$$

The photoemission lines are a convolution of Gaussian and Lorentzian functions. For the quantitative analysis photoelectron spectra have been deconvoluted using Voigt type function in fit macros with IGOR software.

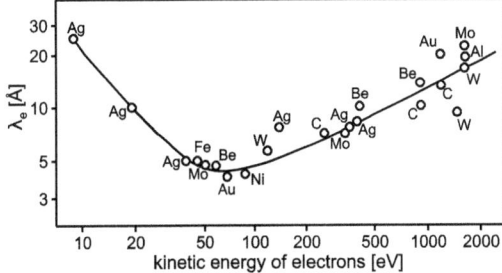

Figure 3.2: The inelastic mean free path of electrons (IMFP) λ_e, which is the surface sensitivity of Photoelectron Spectroscopy is depending on the kinetic energy of electrons. This relation is also called "bathtube curve" (from [238]).

By variation of the excitation energy $h\nu$ the kinetic energy of the photoelectrons is changed. The inelastic mean free path of electrons (IMFP) λ_e is defined as the distance perpendicular to the surface, on which the electron intensity is damped to $\frac{1}{e}$. The IMPF is depending on the kinetic energy of electrons as being shown in the so called "bathtube curve" (Fig.3.2). The IMFP λ_e is almost independent of materials and ranges from 5 - 30 Å, which amounts to 2 − 15 atomic layers. According to the used source, which have different radiation wavelengths, Photoelectron Spectroscopy is divided into Ultraviolet Photoemission Spectroscopy ($h\nu < 41$ eV) and X-ray Photoemission Spectroscopy ($h\nu = 41$ eV - 1500 eV). For Synchrotron Induced Photoemission Spectroscopy synchrotron radiation is needed, which allows a continuous variation of the incident photon wavelength. In this work mainly synchrotron induced Photoelectron Spectroscopy has been applied at two beamlines, the U49-2/PGM2 ($h\nu = 80$ eV - 1500 eV) for core level spectroscopy and the TGM-7 ($h\nu = 14$ eV - 120 eV) for valence band spectroscopy. Since the number of electrons in the synchrotron storage ring decreases with time, the intensity of the incident photons decreases as well. Therefore the spectra were normalized to the beam current of the synchrotron ring to gain calibrated data. In some cases the intensity is referred to a certain electronic level, if attenuation effects due to damping by adsorbates must be eliminated.

Besides photoelectrons also *Auger electrons* are emitted. Auger electrons are created by the relaxation of an electron of outer levels into the primarily created photohole. The energy difference is transferred radiationlessly to another electron, which is leaving the atom. The Auger electron has a constant kinetic energy, thus being independent of the chosen incident excitation energy.

Element analysis and oxidation state Since all elements have defined emission energies, Photoelectron Spectroscopy is suitable for chemical analysis of samples, i.e. for analyzing elements and their concentration on the sample surface. Moreover with Photoelectron Spectroscopy different oxidation states can be distinguished by means of a shift in binding energy of core levels in comparison to the pure element (chemical shift). The binding energy depends on the density of valence electrons around the analyzed atom/ion. The valence electron density changes due to neighbor atoms with different electronegativity, thus the binding energy of the measured core level electrons changes. A shift towards higher binding energy indicates an oxidation, whereas lower binding energy indicates a reduction of the respective element. The chemical shifts of core level lines in different compounds are listed in the literature [239, 240].

Relative concentrations of elements The quantitative composition of samples can be determined from the integral intensity of the photoemission lines after subtraction of the secondary electrons background. In this work the secondary electron background has been subtracted by a *Shirley function* [241]. The intensity of a photoemission line depends on the atomic concentration of the respective element. In addition, the geometry of the spectrometer strongly affects the photoemission intensities, but all these parameters can be summarized in a constant. The intensity of a photoemission line I_A for the element A is thus given by

$$I_A \propto \sigma_A \int_z N_A(x,y,z) \exp\left\{\frac{-z}{\lambda_e \cos\Theta}\right\} dz \qquad (3.6)$$

with σ_A being the cross section of photoionization, N_A the density distribution of atoms of element A, z the coordinate perpendicular to the sample surface, λ_e the inelastic mean free path of electrons (IMFP), and Θ the angle between the normal of the sample surface and the analyzer. Atomic sensitivity factors (ASF), which are listed in tables for each element for an excitation energy of $h\nu = 1486.6$ eV (Al K$_\alpha$ source) [242], include analyzer parameters as the transmission function T, the spectrometer sensitivity and sample parameter like the inelastic mean free path of electrons λ_e and the ionization cross section σ_A. The ASF are different for each type of analyzer. In this work the analyzer *Phoibos 150 MCD-9* from *Specs* has been used. Since experiments have been conducted in this work with different photon excitation energies $h\nu$, the ASF varies due to a change of the ionization cross section σ_A, the IMFP λ_e and the transmission function $T \propto \frac{1}{\sqrt{E_{kin}}}$:

$$ASF \propto \sigma_A \lambda_e \frac{1}{\sqrt{E_{kin}}} \qquad (3.7)$$

3.1 Photoelectron Spectroscopy (PES)

By using the ASF, for homogeneously distributed elements A and B, the relative concentrations are given by

$$\frac{n_A}{n_B} = \frac{I_A \cdot ASF_B}{I_B \cdot ASF_A} \tag{3.8}$$

where n is the concentration of the respective element and I the intensity of the photoemission line.

Thickness of adsorbate layers By means of Photoelectron Spectroscopy the thicknesses of adsorbate layers can be derived from the attenuation of the substrate emission with coverage by the adsorbate. Depending on the thickness of the adsorbate layer, photoemitted electrons have to travel a longer distance to the sample surface and the intensity of the electrons is exponentially decreased. The thickness of the adsorbate layer d_A is

$$d_A = \lambda_e \cdot \ln\left(\frac{I_0}{I_A}\right) \tag{3.9}$$

with I_A as the intensity of the substrate emission line with adsorbate, I_0 as the intensity of the uncovered substrate and λ_e as the inelastic mean free path of electrons in the adsorbate layer.

Resonant Photoemission Spectroscopy (ResPES) ResPES [243–247] is suitable for the investigation of the atomic origin of valence band states. Also it is useful for enhancing the intensity of usually weak features, e.g. the Ti 3d states of TiO_2 within the gap. The PES resonant behavior of anatase [131, 134] and rutile single crystals [132, 248, 249] were described in detail in literature. In PES different ionization processes may occur as displayed in Fig. 3.3.

In the case of TiO_2 the direct photoemission process can be described by

$$3p^6 3d^n + h\nu \longrightarrow 3p^6 3d^{n-1} + e_f^- \tag{3.10}$$

For resonant effects in PES, the excitation energy has to be scanned across the transition of a core level to a valence level state, e.g. $Ti3p \rightarrow Ti3d$ (47 eV) or $Ti2p \rightarrow Ti3d$ (464 eV). This transition leads to the autoionization process, i.e. an electron is excited to an unoccupied Ti 3d state (or Ti 4s state), which relaxes back to its ground state. The energy difference involved in the relaxation process is transferred to the valence electrons and consequently its emission intensity may be enhanced. The binding energies of the involved atomic orbitals are Ti 3p =37.9 eV, Ti 3d =1.3 eV and Ti 2p =459.6 eV, resulting in energy differences $\Delta E_{Ti3p-Ti3d} = 36.6 eV$ and $\Delta E_{Ti2p-Ti3d} = 458.3 eV$. The energy values for both

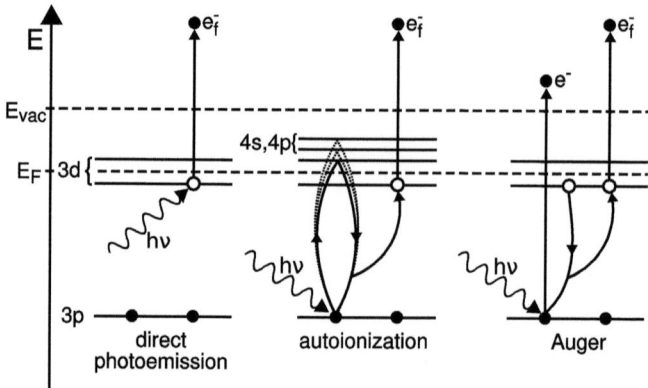

Figure 3.3: Scheme of three different ionization processes during the photoemission spectroscopy. The direct photoemission and the autoionization have identical final states, whereas the *Auger process* causes a double ionization of the atom.

resonances exceed these energy differences. A possible explanation for this mismatch is the excitation of electron into higher energy states.

Thus the relaxing electron transfers its energy radiationless to another electron in the Ti 3d state, which is emitted:

$$3p^6 3d^n + h\nu \longrightarrow [3p^5 3d^{n+1}]^* \longrightarrow 3p^6 3d^{n-1} + e_f^- \quad (3.11)$$

Due to the excitation into different electronic states depending on the excitation energy, the kinetic energy of the emitted electrons varies. For direct photoemission as well as autoionization processes, the kinetic energy of the emitted electron depends on the excitation energy. Moreover the final states in both cases are identical, hence both processes interfere. This leads to an increase of the photoionization cross section and explains the resonant behavior of Ti 3d states at the transition energy.

A different final state is reached by the *Auger process*:

$$3p^6 3d^n + h\nu \longrightarrow [3p^5 3d^n]^* + e^- \longrightarrow 3p^6 3d^{n-2} + e_f^- \quad (3.12)$$

Two electrons are emitted during this process. Since an electron relaxes from the occupied Ti 3d into the 3p state, the transferred energy is constant. The electron emitted from the 3d state has a constant kinetic energy, therefore this process is incoherent with the direct photoemission and does not lead to a resonant behavior.

Due to hybridization of Ti 3d and O 2p states in TiO_2 the resonance occurs via an interatomic process. In this case, the resonant excitation process initially follows Eqn.3.11 but the decay channel is via the O 2p states:

$$[Ti3p^53d^{n+1}]^* + O2p^6 \longrightarrow Ti3p^63d^n + O2p^5 + e_f^- \qquad (3.13)$$

This interatomic resonant excitation allows assumptions about the different features and their bonding nature in the valence band [131, 134].

3.2 Raman Spectroscopy

Raman Spectroscopy relies on the Raman effect (discovered by Raman in 1928 [250]), which describes the inelastic scattering of monochromatic light in the visible, near IR or near UV region. The incident light interacts with vibrational modes in the material. As a consequence besides the elastic scattering (no change in frequency) a shift both towards higher or lower frequency of the incident light is observed, involving the (de)excitation of vibrations. This can explained by means of Figure 3.4.

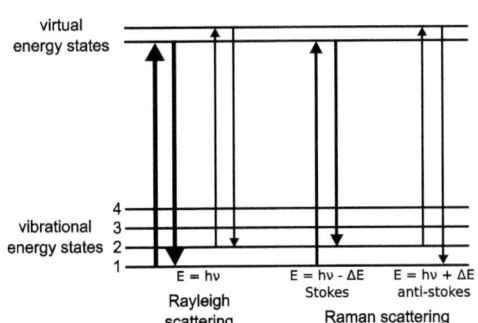

Figure 3.4: Besides the elastic scattering of light (Rayleigh) inelastic scattering (Raman) results in shift towards lower (Stokes) or higher (anti-Stokes) frequency. The thickness of the arrows indicate the intensity of the respective transition

The excitation occurs from a vibrational ground states to a virtual energy state. In the case of the Rayleigh scattering, light is scattered elastically, whereas the Raman scattering is an inelastic scattering. Due to Raman scattering the incident light gains or looses energy. The phonons are relaxing back either to a higher or lower final vibrational state. In the case of the Stokes shift the phonons gain energy, whereas the anti-Stokes shift includes an energy loss of the phonons from a higher initial vibrational energy state. These shifts give information about the phonon modes, which are characteristic for each material. The intensity of the Raman mode depends on the change of polarizability with the normal mode of vibration.

3.3 Grazing Incidence X-Ray Diffraction (GIXRD)

Monochrome X-ray photons are diffracted on crystal planes according to *Bragg's law* [251]

$$n \cdot \lambda = 2d \cdot \sin \Theta \qquad (3.14)$$

where λ is the wavelength of incident X-ray photons, n is a whole number, d is the distance between the crystal planes and Θ is the diffraction angle of the X-rays on the crystal plane. In contrast to the commonly used *Bragg-Brentano geometry* the GIXRD has a constant incident angle ω of X-ray photons, which amounts usually only between 0.5° to 2°. Thus only the detector scans along the 2θ angle. By this method not only reflexes originating from crystallographic planes parallel to the surface are obtained, but also planes in any direction, which fulfill the Bragg reflection condition (Equation 3.5).

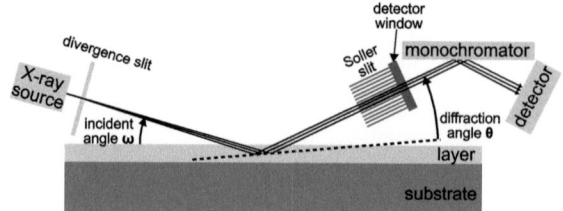

Figure 3.5: The experimental setup of the GIXRD.

Because of the small incident angle this X-ray diffraction method is suitable for phase characterization of thin layers. Only crystal phases can be determined, but a determination of film texture is not possible. The depth information of different reflexes is constant, whereas in the *Bragg-Brentano geometry* it varies, since the penetration depth of the incident x-rays depends on the incident angle.

3.4 Atomic Force Microscopy (AFM)

In 1986 the Atomic Force Microscope (AFM) was invented and introduced by Binnig et al. [252]. It is a further development based on the Scanning Tunneling Microscope (STM). It overcomes the constraint of the Scanning Tunneling Microscope (STM) of being applicable only to conducting samples. As signal the force between a tip and the sample is measured. A schematical picture of the experimental setup is shown in Figure 3.6 a).

As a probe a sharp pyramidal tip (e.g. made out of silicon, Si_3N_4, or diamond) is used, which is located on a flexible cantilever (Figure 3.6 b)). The cantilever is a very thin flat spring, which is scanned over the sample. The vertical deflection and torsion of the cantilever can be detected by measuring the deflection of a laser beam.

3.4 Atomic Force Microscopy (AFM)

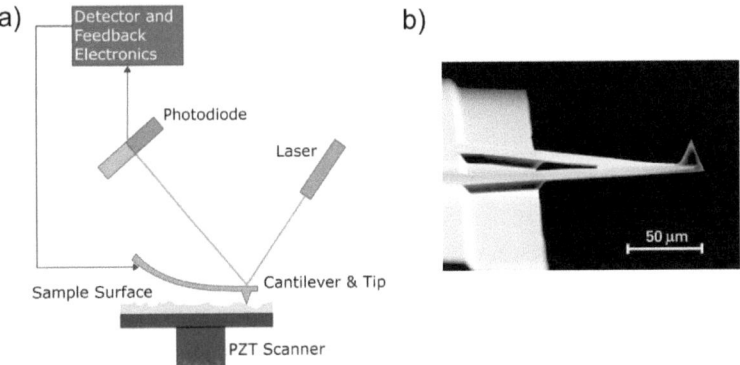

Figure 3.6: The working principle of the Atomic Force Microscope. **a)** The principle assembly [253]. **b)** SEM-image of a tip [254].

The force potential of the tip close to the surface can be described by the *Lennard-Jones potential* [255] $\phi(r)$

$$\phi(r) = -\frac{A}{r^6} + \frac{B}{r^{12}} \qquad (3.15)$$

with A and B being positive constants and r the distance of the tip from the sample. If the tip is in direct contact to the sample in the repulsive range, it is in the so called "contact mode". If the tip just "feels" the attractive forces, the "non-contact mode" is used (Figure 3.7).

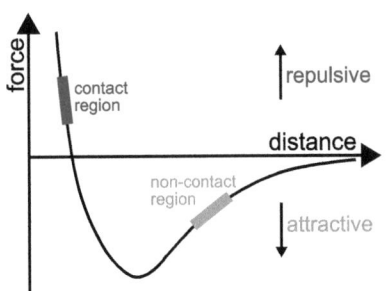

Figure 3.7: The *Lennard-Jones potential*. In the "contact mode" the repulsive force is probed, while with increasing distance the attractive force ("non-contact mode") is probed [254].

The interatomic forces between tip and sample are divided into long and short range contributions. In vacuum long range *van der Waals forces* cover a range up to 100 nm, whereas chemical forces are constrained to few Ångström. The repulsive forces are caused by the forbidden interpenetration of electron orbitals, which is based upon the *Pauli principle*. In air there act additional forces between tip and sample due to adsorbates (e.g. water) on the surface.

In this work only the contact mode (c-AFM) was used. The image mapping in this mode is determined by the spring constant k of the cantilever by the relation [256]

$$k = \frac{E \cdot b \cdot d^3}{4 l^3} \qquad (3.16)$$

denoting E as the Young's modulus, and b, d, and l as the width, thickness and length of the cantilever. The spring constant k has to be in the range of 0.01 - 5 N/m in order to allow a deflection of the cantilever, if being exposed to smallest forces. By the deflection $\Delta = \frac{F_{ts}}{k}$ with F_{ts} being the force between tip and sample, the reflection direction of the laser beam is changed and can be detected on a location sensitive CCD-chip (4 segment detector). Based upon this information a topographic image can be calculated. The Atomic Force Microscope has no real atomic resolution since the tip at its end is much bigger than one single atom. In addition to the vertical deflection of the cantilever a torsion can occur, which provides information about the chemical environment and surface roughness. Mapping of this signal results in lateral force image.

3.5 Scanning Electron Microscopy (SEM)

The Scanning Electron Microscope (SEM) was first introduced by von Ardenne in 1938 [257]. It has a much higher resolution than the conventional light microscopy and can reach magnifications of up to 10^6.

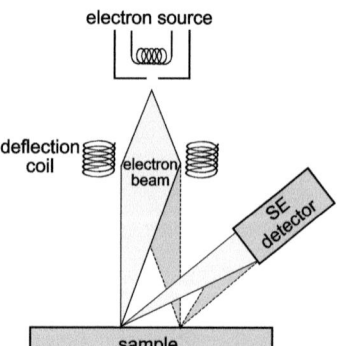

Figure 3.8: The working principle of the Scanning Electron Microscope (SEM).

Electrons are emitted from a tungsten or lanthanum hexaboride (LaB_6) electron cathode, either thermionically upon heating or by field emission due to a high electric field. The electrons are accelerated to the sample. In order to scan the surface, coils are used to change the electric field in a defined way that the electron beam is deflected line by line over the sample surface. Basically three kinds of signals can be detected by an SEM: secondary electrons (SE) due to inelastic scattering, elastically backscattered electrons (BSE) and X-ray photons. The first and most commonly used SE-signal produces three dimensional

3.5 Scanning Electron Microscopy (SEM)

pictures of the sample surface, because the secondary electrons have only a small kinetic energy ($< 50\,\text{eV}$). Thus secondary electrons originate close to the surface. Backscattered electrons (BSE) and X ray photons originate from deeper areas of the sample and can both be used to measure the distribution of elements within the sample. In this work only secondary electron pictures have been taken. As a prerequisite, the samples have to be conductive, which can be accomplished for insulating samples by sputtering a conductive material onto the surface. In this work, the samples have been covered with carbon.

4 Substrate and interface preparation

In this chapter a short description of the sample preparation techniques are given. For the substrate preparation both a sol-gel and a MOCVD process have been applied. The dye and electrolyte adsorption from solution and also the solvent adsorption from the gas phase were performed at the Solid Liquid Analysis System (SoLiAS) at the Berliner Elektronenspeicherring-Gesellschaft für Synchrotronstrahlung (BESSY).

4.1 TiO$_2$ substrate preparation

The sol-gel preparation route is in general applied for manufacturing DSSC devices. In order to reduce contaminations from ambient air, an in-situ method has been applied using MOCVD-preparation.

4.1.1 Sol-Gel preparation

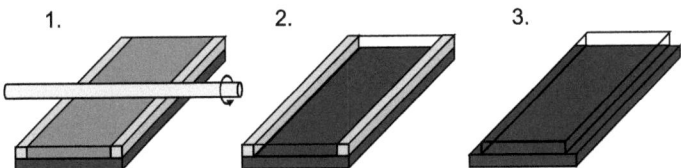

Figure 4.1: The sol-gel process to produce the nanocrystalline anatase. **1.** The anatase slurry is deposited by a glass rod on top of the TCO glass between the scotch tape spacers. **2.** The slurry is dried for 10 minutes in air and turns from white opaque to transparent. **3.** For drying in an oven and the sintering process the scotch tape spacers are removed and the anatase layer is sintered for 30 min at 450 °C in air.

Nanocrystalline and nanoporous TiO$_2$ layers were produced applying the sol-gel process (Fig.4.1) as described by Solaronix S.A. [258]. A white slurry is used, containing 20 % wt. of nanocrystalline anatase powder (~13 nm grain size) in an ethanol solution containing water and polyethylene dioxide (Solaronix Ti-Nanoxide T-series). The white paste has been spread out with a glass rod between two scotch tape spacers. The white slurry turned transparent after drying of 10 min in air, then it was put in a drying oven at 100 °C for 10 min. After that the sample has been taken to an oven, where it was heated to 450 °C within 10 min, then kept at that temperature for 30 min and cooled down to room temperature within 30 min.

4.1.2 Metal Organic Chemical Vapor Deposition (MOCVD)

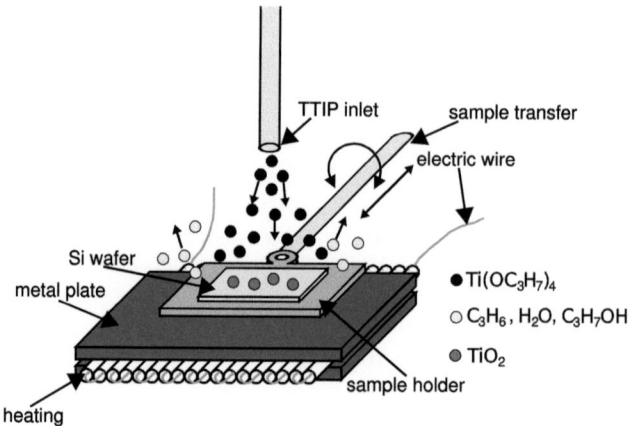

Figure 4.2: The MOCVD process using a custom made heating and TTIP source.

The aim of the MOCVD in situ preparation (Fig.4.2) of anatase layers was to reduce contamination of the TiO_2 substrate. The polycrystalline thin film anatase substrate was prepared in UHV by metal organic chemical vapor deposition (MOCVD) [259, 260] using the single precursor titanium tetra isopropoxide (TTIP) (99.995 % purity, Alfa Aesar). The single precursor inside the glass vessel was heated to 80 °C due to its low vapor pressure. The deposition was performed at a pressure of 10^{-5} mbar in the vacuum chamber (mostly for 2.5 h) onto the native oxide of an Si (111) wafer, which was preheated 30 min before and kept during deposition at 450 °C. During the deposition, the single precursor decomposes on the heated substrate according to the chemical reaction:

$$Ti(OC_3H_7)_4 \longrightarrow TiO_2 + 4C_3H_6 + 2H_2O \qquad (4.1)$$
$$Ti(OC_3H_7)_4 \longrightarrow TiO_2 + 2C_3H_7OH + 2C_3H_6 \qquad (4.2)$$

The two possible reaction paths [261], lead to TiO_2 and to the byproducts propene (C_3H_6) and water (H_2O) or isopropanol (C_3H_7OH). After deposition the sample was kept for 30 min in oxygen ($p_{O_2} = 10^{-5}$ mbar) while cooling down.

4.2 Interface preparation

The preparation of the TiO_2Ru dye/electrolyte interface was done in two stages. The adsorption of the dye and the electrolyte was accomplished under normal pressure conditions,

4.2 Interface preparation

but without contact to ambient air. The (co)adsorption of the solvent was performed from the gas phase in the UHV chamber, which requires a specially designed experimental setup described in the following.

4.2.1 Solid-Liquid Analysis System

The main part of the photoelectron spectroscopy analysis has been conducted at the SoLiAS experimental station, shown in Fig.4.3 [262], which was specifically designed for the analysis of solid / liquid interfaces located at BESSY at the beamlines U49/PGM2 and TGM7. The SoLiAS setup provided a base pressure in the range of 10^{-10} mbar. The SoLiAS setup (Fig. 4.3) can be divided into three levels: The first level is the analysis level, where Photoelectron Spectroscopy has been performed (Sec.3.1). For taking the photoelectron spectra, a hemispherical analyzer (model "Phoibos" from SPECS GmbH) has been used. All spectra taken were referenced to the gold Fermi edge with respect to binding energy of electrons. The second level is the preparation level for MOCVD in situ preparation or subsequent annealing steps of substrate material (Sec.4.1.1 and Sec.4.1.2). The third level offers the opportunity to prepare solid / liquid interfaces, which is the unique feature of SoLiAS (Sec.4.2.2 and Sec.4.2.3). Here the dye / electrolyte and solvent adsorption were performed.

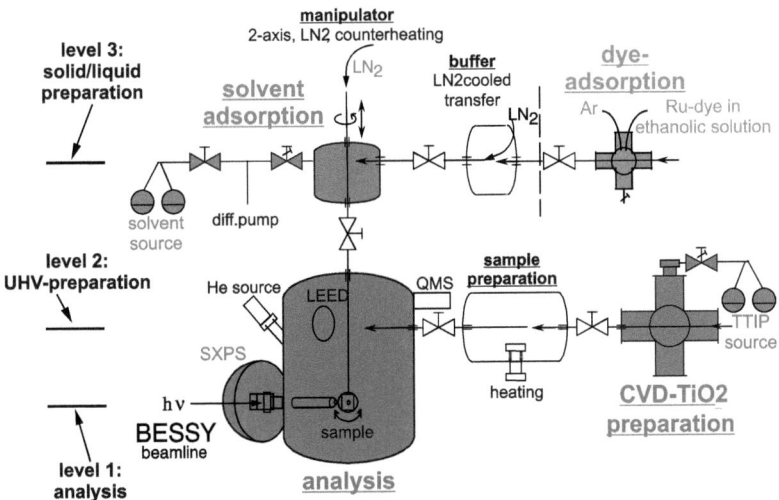

Figure 4.3: The Solid Liquid Analysis System (SoLiAS) experimental station.

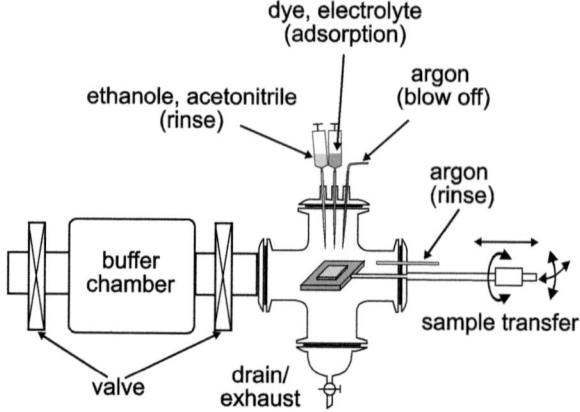

Figure 4.4: Glass cell

4.2.2 Dye and electrolyte adsorption

To adsorb the dye (bought from Solaronix S.A. Switzerland, dissolved in ethanol solution, 20 mg/100 ml) and/or the electrolyte (LiI or the molten salt, 1-propyl-3-methylimidazolium iodide (PMII), dissolved in acetonitrile) an UHV-integrated glass cell was used (Figure 4.4). This glass cell is attached directly to the SoLiAS station and continuously purged with pure argon continuously allowing a direct transfer to the ultra high vacuum (UHV) by avoiding contact to ambient air. The dye Ru^{II}(2,2'-bipyridine-4,4'-dicarboxylate)$_2$(NCS)$_2$ was adsorbed for 3 min and afterwards the TiO$_2$ film was rinsed with pure ethanol in order to achieve monolayer coverage. An argon pipe was used to blow off the residual solvent after rinsing and to dry the sample. The electrolyte adsorption has been accomplished by exposing the TiO$_2$ film to the solution and by waiting for complete evaporation of the solvent. But the sample has not been rinsed before introducing into UHV.

4.2.3 Solvent (co)adsorption

For coadsorption of solvents acetonitrile was used (available in argon filled resealable Chem-Seal bottles with 99.8 % purity, Alfa Aesar). A glass vessel was filled with acetonitrile inside a glove box in argon atmosphere. This vessel has subsequently been attached to the UHV system via a leak valve. As an additional solvent the unpolar benzene with a purity >99 % was investigated and treated in an analogous way. In order to remove argon and gaseous contaminations out of the glass vessel, the vessel was pumped down to HV after the solvent was freezed with liquid N$_2$. Coadsorption of the solvents were accomplished from the gas phase onto the sample, which was cooled by a liquid nitrogen (LN$_2$) reservoir inside the manipulator (Fig.4.5). The solvent source has been checked for purity with a quadrapole

4.2 Interface preparation

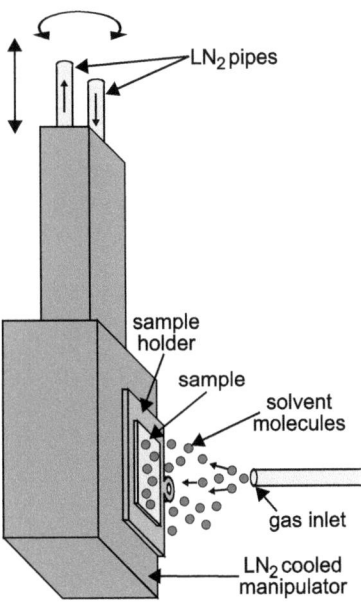

Figure 4.5: Gas phase adsorption

mass spectrometer. As the last step in (co)adsorption experiments the desorption of volatile species has been performed at room temperature for several minutes to few hours.

Part III

Results and Discussion

5 Scope of this work

The aim of the present work is to obtain a comprehensive understanding of the complex electronic interactions operative at the working electrode of the Dye Sensitized Solar Cell. The focus lies on the electronic coupling of TiO_2 substrate, dye, electrolyte salt and solvent. In contrast to p/n junction solar cells the working principle of the Dye Sensitized Solar Cell relies on interfacial charge transfer involving chemically different components. Since the interface dimension is in the range of one dye monolayer, high surface sensitivity of the analysis is mandatory. Thus Synchrotron Induced Photoemission Spectroscopy has to be applied making use of its high surface sensitivity. The characterization of each component has to be performed in order to be able to attribute the respective photoemission lines appropriately.

Since the TiO_2 substrate has been produced in two different ways, on one hand by Sol-Gel preparation (ex situ) and on the other hand by MOCVD (in situ), it is necessary to compare both substrate types. The characterization involves morphology (SEM) and crystalline phases like either anatase or rutile (GIXRD, Raman), and electronic structure (SXPS).

As assumed by several groups [48, 76, 78, 144, 145], defect states, which are attributed as surface states [138], affect the device performance negatively by acting as recombination centers after trapping the photoinjected electrons. On the other hand, Gregg et al. [17, 80] claim, that upon UV illumination defects are created, which are beneficial for the performance of the DSSC in terms of providing a conduction path for electrons. One aim of this Photoelectron Spectroscopy study is to investigate, how these defect states can be influenced. Since it is assumed, that these trap states arise also from oxygen vacancy states at the substrate surface (Section 2.1.4), the behavior of these defects has to be investigated in the presence of all involved adsorbates at the working electrode.

All charge transfer reactions depend on the alignment of the electronic energy levels of substrate, dye and electrolyte. Therefore it is attempted to investigate the position of all relevant energy levels of the involved components by means of Photoelectron Spectroscopy. By coadsorption experiments of dye, electrolyte salt and solvent, the electronic alignment on the working electrode needs to be elucidated.

The key of the Dye Sensitized Solar Cell concerning device operation is the dye molecule and its orientation relative to the TiO_2 substrate surface. In order to enhance the dye molecule rereduction (Section 1.3.2) and to minimize the detrimental back electron transfer (Section 1.3.4.2), the adsorption geometry on the substrate surface needs to be elucidated.

6 The TiO$_2$ substrate

In this work the anatase substrate was made by two different production techniques – the Sol–Gel method and the MOCVD process (Section 4.1.1 and 4.1.2). In the following the substrates are referred as nc-TiO$_2$ for the ex-situ produced Sol-Gel TiO$_2$ and as CVD-TiO$_2$ for the in-situ prepared MOCVD layers. Both substrates have been investigated with respect to their crystal phases by Raman Spectroscopy and GIXRD, their morphology by SEM and AFM and their electronic properties by SXPS and ResPES.

6.1 Crystalline phases

The TiO$_2$ substrate in the DSSC acts as the electron conducting material. Hence the conductivity, which is determined by the mobility of the charge carriers, is an important property of the material. As mentioned in Section 2.1.1, the electron mobility is one magnitude higher for anatase compared to rutile. Thus an aim is to produce anatase substrates to enhance electron conductivity, which is crucial for the performance of the DSSC device.

6.1.1 GIXRD

In order to determine the phase purity of both nc-TiO$_2$ and CVD-TiO$_2$, X-ray diffraction experiments with grazing incidence (GIXRD) have been performed (Figure 6.1). The Sol-Gel slurry consists of TiO$_2$ particles being at least 70 % anatase and less than 30 % rutile as specified by the manufacturing company Solaronix S.A. In contrast the situation for the MOCVD-TiO$_2$ is not clear, thus it is of special interest to investigate its crystalline phase.

In both cases the crystalline phase could be unambiguously determined being anatase by means of its characteristic reflexes. No contributions of rutile could be detected. The nc-TiO$_2$ shows in addition reflexes originating from the SnO$_2$:F layer, which could be caused by an incomplete coverage of the SnO$_2$:F layer by the nc-TiO$_2$ layer. Since the incident angle of the X-ray beam is fixed in the GIXRD geometry, planes of different angles relative to the surface plane fulfill the reflection condiction. This implies that no information on texture is available. On the other hand it was not possible to accomplish X-Ray Diffraction with the *Bragg-Brentano* geometry, because both types of TiO$_2$ layers have an insufficient thickness of only around 1 μm.

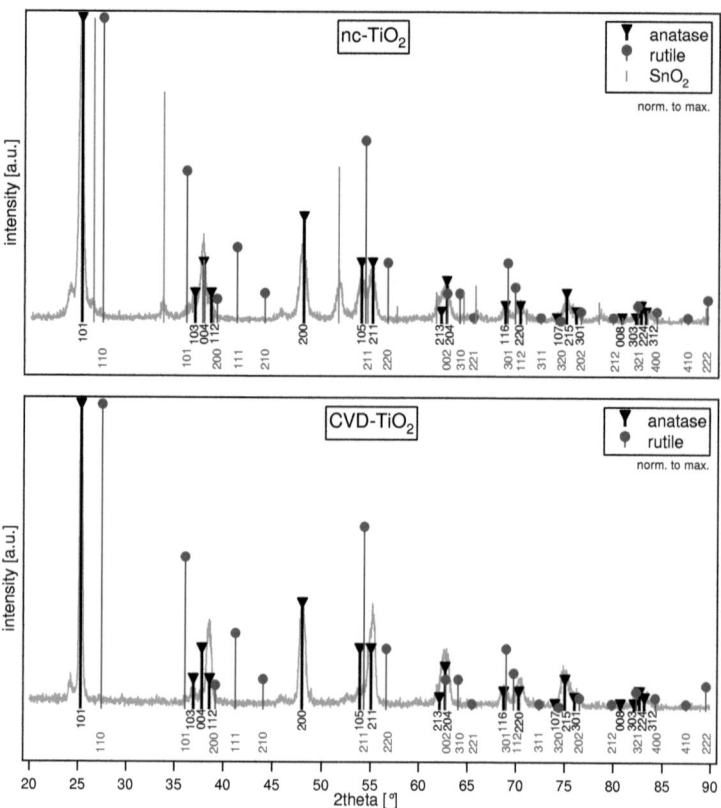

Figure 6.1: GIXRD measurements of the produced TiO$_2$ substrate measured with an incident angle of 2°. Below each diffraction pattern the *Miller Indices* (hkl) are given. **top:** The Sol–Gel layers sintered at 450 °для 30 min. **bottom:** The TiO$_2$ substrate deposited by MOCVD.

6.1 Crystalline phases

6.1.2 Raman

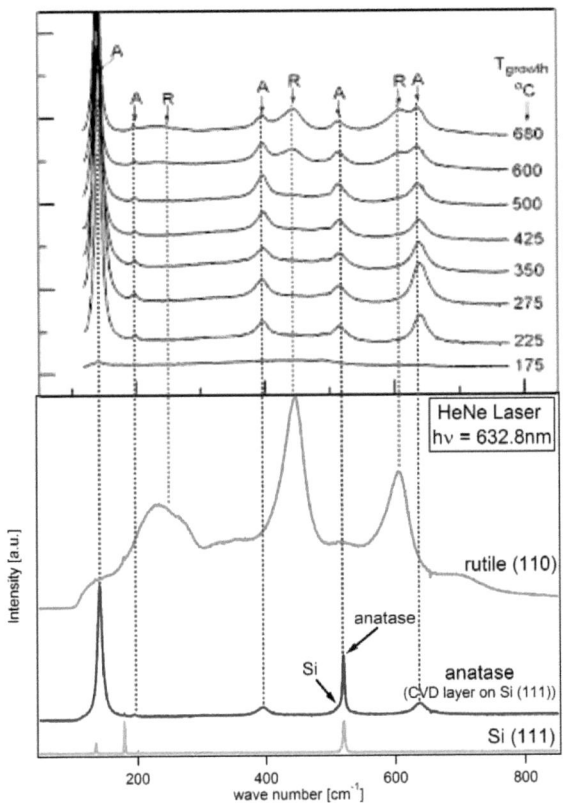

Figure 6.2: Raman spectra of CVD anatase thin films and rutile (110) crystal compared with a spectra series taken from literature [263], that shows the transformation of anatase A to rutile R with temperature.

Also Raman measurements have been performed to give additional evidence for the crystalline phase of the CVD-TiO$_2$. A microfocal Raman spectrometer with a HeNe-Laser ($\nu = 632.8$ nm) was used. Raman spectra of single crystalline rutile (110) and CVD-TiO$_2$ have been taken, which are displayed in Figure 6.2. The comparison of these spectra with literature data [263] clearly evidences, that the CVD-TiO$_2$ contains only the anatase phase.

6.2 Morphology

Both nc-TiO$_2$ and the CVD-TiO$_2$ layers have been investigated by SEM and AFM with respect to their morphology. Size and shape of the TiO$_2$ particles as well as microstructure and thickness of the layer were of interest.

6.2.1 nc-TiO$_2$

6.2.1.1 SEM

The top view SEM images (Fig.6.3 a)) show the crystal size and shape of the TiO$_2$ particles. The TiO$_2$ nanoparticles consist of crystallites with rounded edges and corners connected to each other to form a porous network. The evident particle size amounts to about 50 – 100 nm according to these SEM measurements.

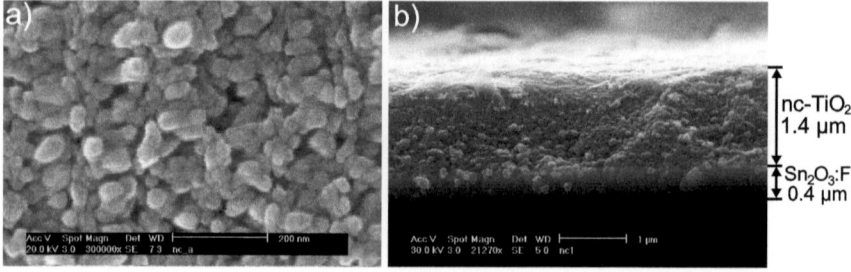

Figure 6.3: SEM micrographs of TiO$_2$ anatase films prepared by applying the sol-gel process (sintered for 30 min at 450 °C in air). **a)** Top view. **b)** Side view of the same layer.

The side view images (Fig.6.3 b)) allow to estimate the thickness and the microstructure of the TiO$_2$ layer. The underlying transparent conducting oxide layer (TCO) measures 0.4 μm, whereas the nc-TiO$_2$ layer has a thickness of 1.4 μm. The microstructure of the nanoporous layer is globular.

6.2.1.2 AFM

The AFM picture (Fig.6.4 left) confirms the porous morphology with globular crystallites as already deduced from the SEM measurements described above. The average particle size is estimated to be about 100 nm, which is of similar size as determined in the SEM experiments. Also a profile over a distance of 5 μm is shown (Fig.6.4 right). From the deepest to the highest point the height difference amounts about 10 nm, which is much less than the average size of one particle. This is most likely due to the size of the AFM tip, which is too big to intrude in between two particles.

6.2 Morphology

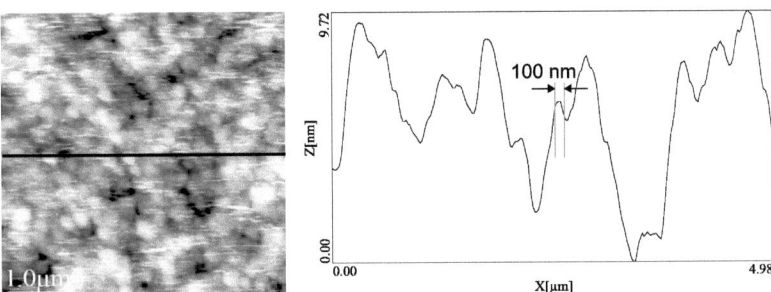

Figure 6.4: left: The AFM image of the nc-TiO$_2$ over an area of 5×5 μm. The black line shows the scan line used for a profile. **right:** The profile across the line. The particle diameter is estimated to be about 100 nm.

6.2.2 MOCVD-TiO$_2$

6.2.2.1 SEM

The top view SEM image (Fig.6.5 a)) shows that the TiO$_2$ particles formed are small plates. The normal of this small plates are oriented almost perpendicular to the underlying surface normal. The plates are stacked parallel in small domains, forming a columnar microstructure. Additionally, there are areas on the CVD-TiO$_2$, which show a granular microstructure as displayed in Figure 6.5 b). In this case the crystallite size varies more, i.e. it ranges from 20 nm up to 200 nm (note the different scale compared to Figure 6.5 a)). Moreover the habitus of anatase single crystals can be identified (white rectangular), which shows one triangular face of an octahedron, i.e. the (101) surface (compare with Fig.2.3 a) and c), page 32).

The side view (Fig.6.5 c)) shows, that in contrast to the nc-TiO$_2$ the CVD-TiO$_2$ has a columnar microstructure. The thickness of the underlying natively oxidized Si(111) layer is around 0.4 μm, and the TiO$_2$ layer is about 0.9 μm.

6.2.2.2 AFM

The AFM image of the MOCVD TiO$_2$ (Fig.6.6) reveals a morphology more similar to the SEM picture shown in Figure 6.5 b) than in Figure 6.5 a). The reason for this differences could be the fact, that the tip of the AFM is not able to resolve the fine crystallite plates as displayed in Figure 6.5 a). The TiO$_2$ particle size of 160 nm is similar as found by SEM. A 5 μm line profile shows a similar roughness as for the nc-TiO$_2$, which amounts to be about 8.5 nm.

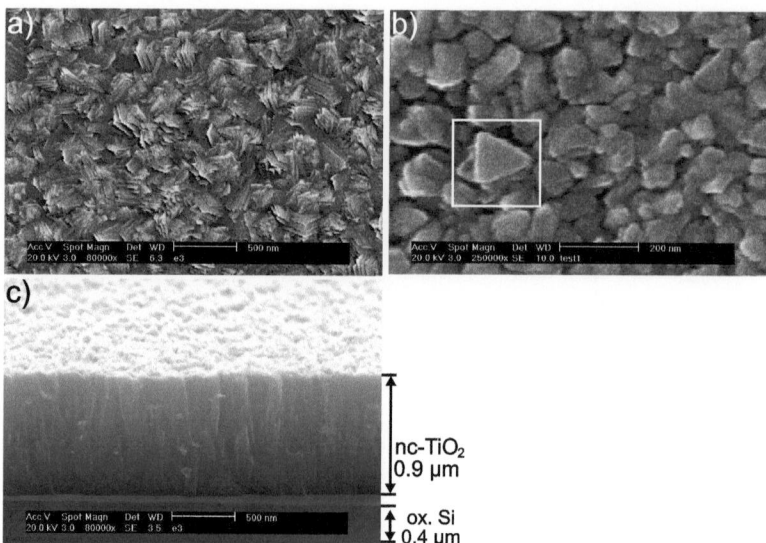

Figure 6.5: SEM micrographs of insitu prepared TiO$_2$ anatase films using MOCVD (2.5 h and 10^{-5} mbar TTIP, substrate heated at 450 °C, 2.5 A heating current; kept after deposition at 450 °C in oxygen). **a)** Top view, showing the particles being small plates. **b)** Top view of another sample. Here the particles are of granular shape. **c)** Side view with a layer thickness of about 1 µm.

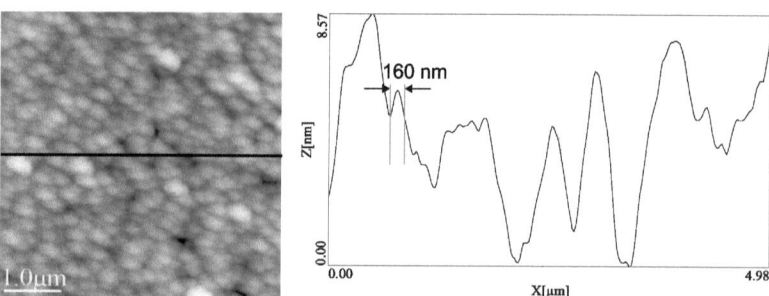

Figure 6.6: left: AFM image of CVD-TiO$_2$ over an area of 5x5 µm. The black line shows the scan line used for a profile. **right:** According to the profile across the line, the particle diameter amounts to be about 160 nm.

6.2.3 Summary

Concerning the anatase particle size our results by GIXRD and Raman experiments slightly deviate from the literature (Section 2.1.1). We have deposited anatase particles of a size ranging from 20 to 200 nm for the CVD-TiO$_2$. In contrast the phase stability landscape displayed on Page 29 (Fig.2.4) predicts anatase being stable only below a particle size of 30 nm. A possible reason might be the different ambient conditions like vacuum instead of ambient pressure. The nc-TiO$_2$ particles show a diameter of 50 nm, although the dispersed particles are specified by Solaronix S.A. being around 13 nm before sintering. Thus the increased particle size of 50 nm is evidently due to the sintering process applied in this work.

Comparing the morphology of both substrates reveals differences: Whereas the nc-TiO$_2$ consists of round shaped crystallites, the SEM pictures of the CVD-TiO$_2$ show either small plate-like crystallites or small crystalline grains revealing the habitus of single crystalline anatase. While the nc-TiO$_2$ substrate consists of a granular microstructure (Fig. 6.3 b)), the CVD-TiO$_2$ side view shows a columnar microstucture (Fig. 6.3 c)). For the device the two types of microstructures may have different consequences: The granular microstructure might favor the contact of the dye sensitized substrate surface with the electrolyte and enhances the light absorption, whereas the columnar microstructure may result in a better charge transport.

Nevertheless the morphology has no relevant influence on the electronic structure of the substrate material, as shown in Section 6.3. Hence for this work it is not of importance, which morphology is present in an experiment.

6.3 Electronic structure and defects

The electronic structure of anatase, especially in form of nanocrystallites, is of high interest for research and application concerning the DSSC. Since Photoelectron Spectroscopy measurements of nanocrystalline anatase is rather seldom, there is an urgent need to perform such experiments. Anatase samples produced using the two presented procedures. Those two materials have been analyzed with respect to changes of the electronic structure (see below).

6.3.1 Survey spectrum of core level emissions

Photoemission spectra taken at $h\nu = 900$ eV (Figure 6.7) give an overview of the core level emissions of both nc-TiO$_2$ and CVD-TiO$_2$. In Table 6.1 the binding energy values of the core level emissions are listed, which were determined from detail spectra of the respective emissions.

The measured binding energy values of the Ti 2p and O 1s level match reasonably with literature data, i.e. the XPS data of Orendorz et al. [140], who measured as well nc-TiO$_2$

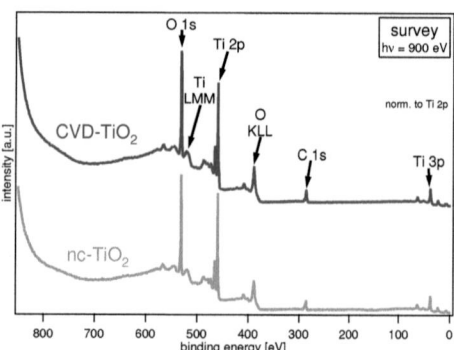

Figure 6.7: Survey spectra of TiO_2 taken at $h\nu = 900\,eV$, which shows the most intense core level emissions.

	O 1s [eV]	Ti $2p_{3/2}$ [eV]	C 1s [eV]	Ti 3p [eV]	O 2p [eV]
$E_{B,exp.}$	530.8	459.5	285.9	37.9	22.8
$E_{B,lit.}$	530.7	459.3			

Table 6.1: The measured binding energies of the core levels of TiO_2 compared with literature data from [140] (Ti 2p and O 1s only).

prepared by the sol-gel process. The carbon C 1s emission demonstrates, that a certain contamination level for both sample types is present, which arises from residuals of carbon containing ingredients of the slurry (nc-TiO_2) or the single precursor TTIP (CVD-TiO_2). Therefore even for the in situ prepared CVD TiO_2 samples a complete elimination of the carbon contamination could not be accomplished.

6.3.2 Surface contributions in core level emission spectra

In order to gain more information about the different elements and their oxidation state core level spectra have been taken from the Ti 2p and O 1s emission. The Ti 2p and the O 1s core level have been measured at different excitation energies, at $h\nu = 600\,eV$, $900\,eV$ and with an Al-K_α X-ray source at $1486.6\,eV$ (Figure 6.8 left top and bottom). Since a variation of the excitation energy leads to a change of depth information, the substrate has been investigated with different surface sensitivities, which enables us to distinguish between surface and bulk contribution of the substrate. Values for the IMFP of electrons are given in Table 6.2.

6.3.2.1 Ti 2p

The Ti 2p doublet consists of the Ti $2p_{3/2}$ ($E_B = 459.5\,eV$) and Ti $2p_{1/2}$ emission ($E_B = 465.2\,eV$), which are split by $5.7\,eV$ [239] due to different spin-orbit coupling. To analyze this core levels in detail, a fitting procedure using Gauss-Lorentz profiles has been applied after a Shirley background substraction. As important information the spectral line shape

6.3 Electronic structure and defects

analysis shows a low binding energy shoulder of the Ti 2p emission relative to the main emission peak, which is addressed as Ti^{3+} states [264, 265]. The existence of Ti^{3+} is generally assigned to oxygen vacancies on the surface [266, 267]. The emission is shifted 1.6 eV to lower binding energies compared to the main emission. We observe a clear decrease of the intensity of the Ti^{3+} states with decreased surface sensitivity, as one may follow in the graph at the right of each set of presented spectra (Fig.6.8 right top and bottom). It is important to note, that no intensity of Ti^{3+} states is observed with the less surface sensitive XPS measurements. These data allow the conclusion that the Ti^{3+} states are located at the surface.

Furthermore, the Ti^{3+}-intensity for nc-TiO_2 and CVD-TiO_2 is slightly different. For CVD TiO_2 a higher concentration is observed, which is most likely caused by the manufacturing process. Unlike the nc-TiO_2 the CVD-TiO_2 was deposited in vacuum, ambient air was excluded. Although titanium tetraisopropoxide (TTIP) is a single source precursor (refer to Section 4.1.2), i.e. both titanium and oxygen are provided as products of the pyrolysis process, the substrate surface seems to be slightly more reduced than it is the case for nc-TiO_2.

For estimating the coverage at the surface by the Ti^{3+} states from the spectra a model described by Mönch [268] was applied. In this model an exponential damping of the photoemitted electron intensity between identical and equally spaced lattice planes is assumed. According to this model, the total intensity of the electrons emitted normal to the surface I_{tot} equals

$$I_{tot} = I_S \sum_{n=1}^{\infty} \exp\left\{-\frac{(n-1)d_{hkl}}{\lambda}\right\} \quad (6.1)$$

$$= I_S \frac{1}{1 - \exp\left\{-\frac{d_{hkl}}{\lambda}\right\}} \quad (6.2)$$

with I_S as the intensity of electrons emitted from the nth plane, n as the number of the plane being located from another plane by the lattice spacing d_{hkl} and λ being the IMFP of emitted electrons in the material. Since we are interested in the fraction of monolayer coverage with Ti^{3+} surface sites, we consider the ratio of I_S/I_{tot}. Hence for the fraction of the surface contribution related to the total intensity we can write

$$\frac{I_S}{I_{tot}} = R_S = 1 - \exp\left\{-\frac{d_{hkl}}{\lambda}\right\} \quad (6.3)$$

The ratio for the surface intensity I_S to the bulk intensity I_B, i.e. the intensities of the layers $n = 1$ divided by the intensity of the layers from $n = 2$ to $n = \infty$, is given by

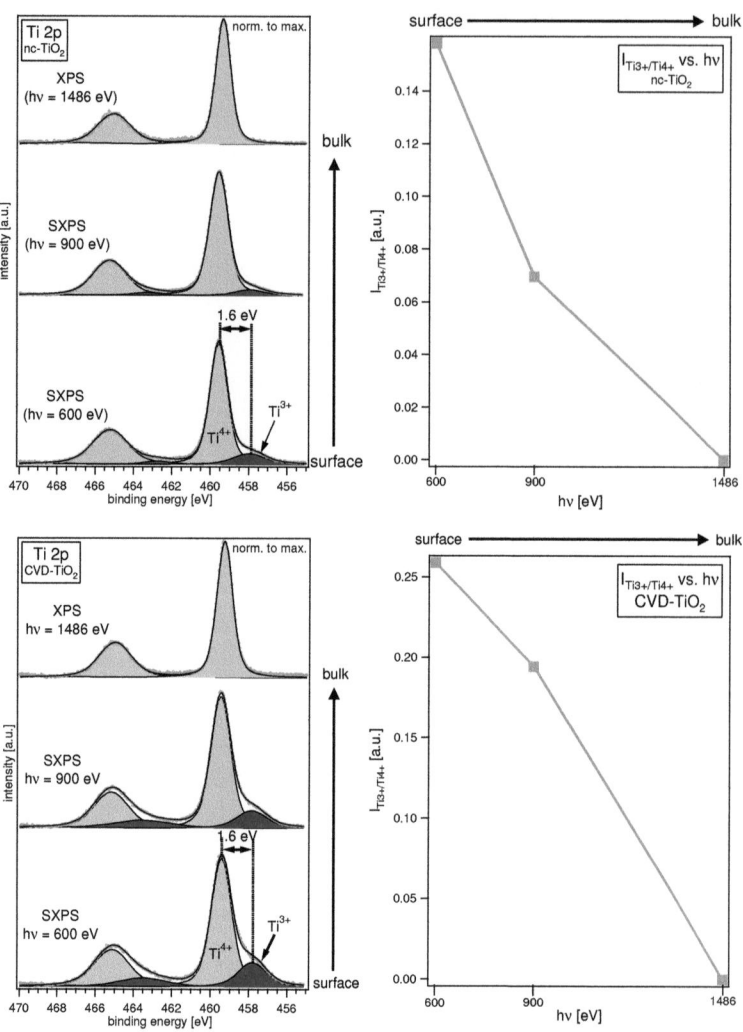

Figure 6.8: The Ti 2p core level emission of the TiO_2 substrate. **left top:** The Ti 2p emission from the nc-TiO_2 substrate fitted with Gauss-Lorentz (Voigt) profiles. **right top:** The related integrated intensity of the Ti^{3+} states relative to the Ti^{4+} states. **left bottom:** The Gauss-Lorentz fit for the Ti 2p emission from the CVD-TiO_2 substrate. **right bottom:** The related integrated intensity of the Ti^{3+} states relative to the Ti^{4+} states.

6.3 Electronic structure and defects

$$\frac{I_S}{I_B} = \frac{R_S}{1-R_S} = \exp\left\{\frac{d_{hkl}}{\lambda} - 1\right\} \qquad (6.4)$$

For the (101) surface the lattice spacing is $d_{101} = 3.52$ Å. In the following Table 6.2 the values for the estimated fraction of monolayer coverage of the Ti^{3+} states $f_{ML}(Ti^{3+}) = \frac{I_{Ti^{3+}/Ti^{4+}}}{I_S/I_B}$ are listed. Also values for the excitation energy $h\nu$, the kinetic energy of the photoemitted electrons E_{kin}, the IMFP λ (from [269]), the surface to bulk intensity ratio $\frac{I_S}{I_B}$ and the intensity ratio of the Ti^{3+} to the Ti^{4+} states $I_{Ti^{3+}/Ti^{4+}}$ are provided.

				nc-TiO$_2$		CVD–TiO$_2$	
$h\nu$ [eV]	E_{kin} [eV]	λ [Å]	I_S/I_B	$I_{Ti^{3+}/Ti^{4+}}$	$f_{ML(Ti^{3+})}$	$I_{Ti^{3+}/Ti^{4+}}$	$f_{ML(Ti^{3+})}$
600	140	6.3	0.75	0.16	0.21	0.26	0.35
900	440	12.0	0.34	0.07	0.21	0.19	0.56
1486	1026	22.1	0.17	0	-	0	-

Table 6.2: The values for the estimated fraction of monolayer coverage of the Ti^{3+} states of both substrates. The IMFP values have been taken from [269].

Under the given assumptions, at the nc-TiO$_2$ 21 % of a monolayer are covered with Ti^{3+} sites, whereas for the CVD-TiO$_2$ the surface concentration ranges from 35 % to 56 % percent. The reason for the big deviation between the two values obtained with different excitation energies for the CVD-TiO$_2$ could be, that the Ti^{3+} states might not be ideally distributed only over the first monolayer of the material as being assumed in this estimation. If the Ti^{3+} states are additionally located below the topmost surface, the intensity of the Ti^{3+} emission will not decay as fast as in the case of being distributed only over the first monolayer.

6.3.2.2 O 1s

The O 1s emission has been taken at the same excitation energies like the Ti 2p core level, at $h\nu = 600$ eV, 900 eV and 1486 eV, which is shown in Figure 6.9. Analogous to the Ti 2p spectra, the aim is to distinguish between bulk and surface components.

The spectra taken at $h\nu = 600$ eV show a strong shoulder at higher binding energy with respect to the main emission. By applying a Voigt profile fitting procedure, the core level emission could be deconvoluted into three components: The main emission line is located at $E_B = 530.8$ eV, which is addressed to bulk oxygen. According to literature data [270] the next component is shifted towards higher binding energy by 0.9 eV ($E_B = 531.7$ eV). This component is allocated to a bridging oxygen [270, 271], i.e. a 2-fold coordinated oxygen (2c-O) at the outermost surface of the material (see page 33 Fig.2.5). The component at highest binding energy ($E_B = 533.2$ eV) is being attributed to chemisorbed OH-groups or alternatively to attached hydrogen atoms [270–272]. Molecular water can adsorb from ambient air (nc-TiO$_2$) or from residuals from the pyrolysis of the single precursor TTIP (CVD-TiO$_2$). But since molecularly adsorbed water is located at $E_B = 534.6$ eV as shown

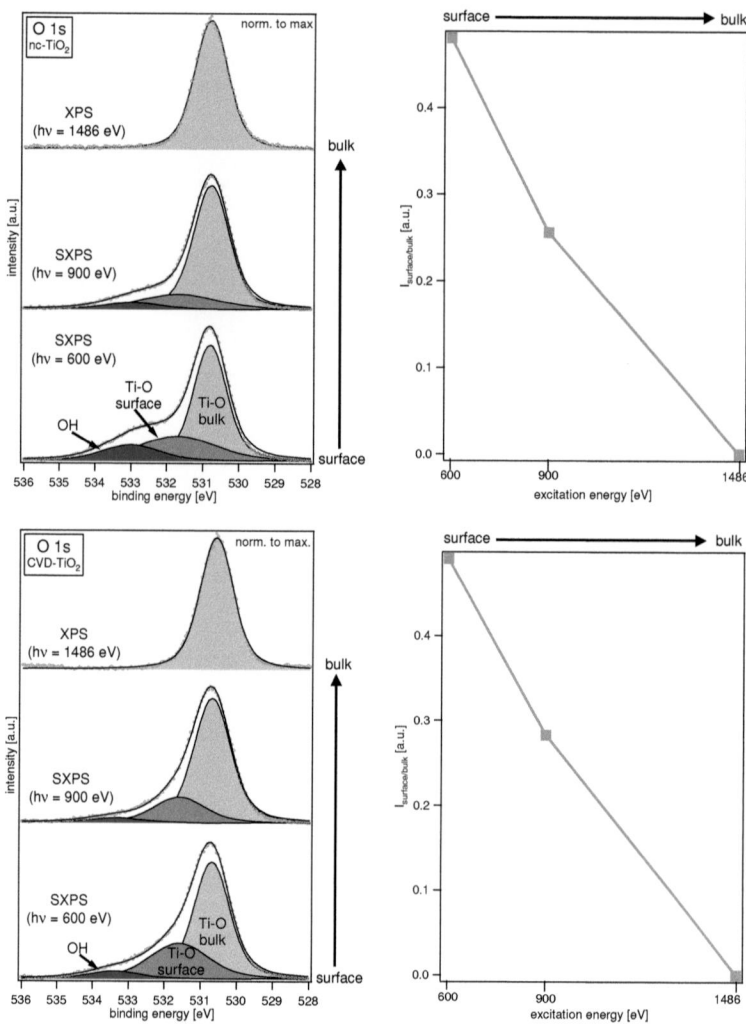

Figure 6.9: The O 1s core level emission of the TiO$_2$ substrate. **left top:** The O 1s emission from the nc-TiO$_2$ substrate fitted with Gauss-Lorentz (Voigt) profiles. **right top:** The related integrated intensity of the surface oxygen relative to the bulk oxygen states. **left bottom:** The Gauss-Lorentz fit for the O 1s emission from the CVD-TiO$_2$ substrate. **right bottom:** The related integrated intensity of the surface oxygen relative to the bulk oxygen states.

6.3 Electronic structure and defects

by adsorption of a thick water layer (compare Figure 7.8),Therefore, it can be excluded to exist on the untreated sample. From the change of intensity with surface sensitivity it is evident that both components at higher binding energy are surface components. For the bridging oxygen, its relative intensity has been estimated using again the model of Mönch (Equation 6.4). The monolayer coverages $I_{Ti-O,s/b}$ are displayed in Table 6.3.

				nc-TiO$_2$		CVD-TiO$_2$	
$h\nu$ [eV]	E_{kin} [eV]	λ [Å]	I_S/I_B	$I_{Ti-O,s/b}$	$f_{ML(Ti-O,b)}$	$I_{Ti-O,s/b}$	$f_{ML(Ti-O,b)}$
600	69	5.4	0.92	0.48	0.52	0.49	0.53
900	369	10.6	0.39	0.26	0.67	0.28	0.71
1486	955	20.9	0.18	0	-	0	-

Table 6.3: The values for the estimated fraction of monolayer coverage of the bridging oxygen (2c-O) on both substrates. The IMFP values have been taken from [269].

Since the ideal anatase (101) surface has 50 % coverage with bridging oxygen, the coverage value estimated by means of the 600 eV spectra matches very well. In contrast the calculated monolayer coverages from the 900 eV spectra exceed the 50 %.

6.3.3 Valence band and band gap

Photoelectron spectra have also been taken with an excitation energy of $h\nu = 90$ eV in order to record the Ti 3p emission line for intensity normalization. In addition, it was possible to measure spectra with high surface sensitivity. Finally the valence band structure is accessible, which reveals strong similarities of both nc- and CVD-TiO$_2$ (Fig.6.10 left).

The titanium and oxygen levels are located at the same binding energies E_B for both kinds of samples: Ti 3p at 37.9 eV and O 2s at 22.8 eV. A strong feature from around $E_B = 10$ eV to 3.6 eV shows valence band emissions mainly originating from O 2p states (see Section 2.1.3 for details). It has to be mentioned, that the additional spectral features of the CVD-TiO$_2$ substrate between 20 eV and 13 eV cannot be assigned until now.

Detailed measurements have been performed for the gap region between 4 eV up to the Fermi level (Fig.6.10 right). For comparison a spectrum of rutile has been added. Taking the leading edge of the valence band maximum (VBM) for both anatase spectra leads to a value of 3.6 eV, whereas the VBM of the rutile single crystal is at 3.0 eV. As one can see all samples show gap states at $E_B = 1.3$ eV for nc-TiO$_2$, 1.2 eV for CVD-TiO$_2$ and 1.1 eV for single crystalline rutile. In literature those gap states are assigned to oxygen vacancies V_O (see Section 2.1.4). The intensity of these gap states correlates with the intensity of the Ti^{3+} states, observed as a low binding energy shoulder of the Ti^{4+} emission (see Section 6.3.2). Additional evidence for the correlation is given by solvent adsorption experiments described in Section 7.

Interestingly we observe in the gap region states just below the Fermi level $Ti3d < E_F$, whose origin is still not clear to date [21]. The intensity ratio of these states relative to the

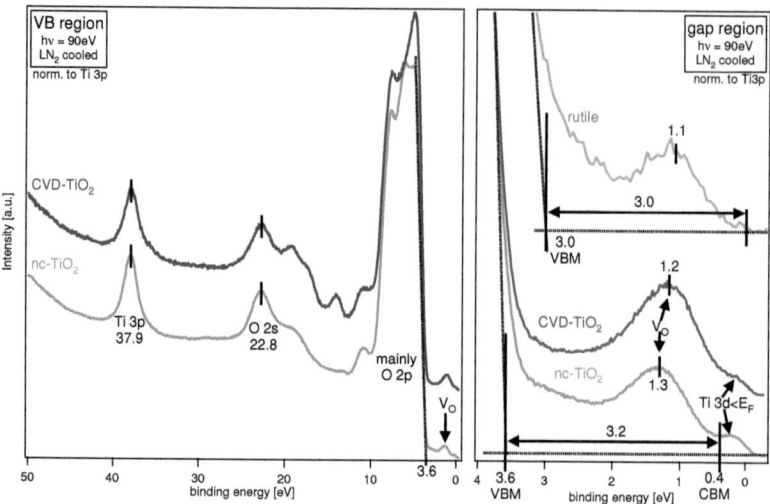

Figure 6.10: left: Valence band spectra of ex-situ prepared nc-TiO$_2$ and in-situ prepared CVD-TiO$_2$. All numbers are binding energy values in eV. **right:** The gap region between valence band maximum and Fermi level, that shows oxygen vacancies V_O and states just below the Fermi level $Ti3d < E_F$. For comparison a single crystal rutile spectrum has been added.

intensity of the oxygen vacancies vary for the anatase samples, being higher for nc-anatase compared to CVD-anatase. In contrast the single crystalline rutile does not show these $Ti3d < E_F$ states at all. In the literature the role of states close to the conduction band is controversively discussed for the performance of the device. Many groups argue, that these states play a detrimental role for the device, because they may act as recombination centers [48, 76, 78, 144, 145]. In contradiction to those groups, Gregg et al. claim that those states act as a delocalized transport band, which enhances electron transport in the anatase material [17, 80]. Very recently, Kitao et al. [273] performed calculations based on our spectroscopic data [265]. They suggest, that the $Ti3d < E_F$ states may correspond to intrinsic acceptor states, which enhance electron transfer from the dye to the TiO$_2$ substrate. However, in literature it is not distinguished clearly between the V_O and the $Ti3d < E_F$ states for discussing their role within the device. A possible solution of this controversy might be, that on one hand the V_O states act as recombination centers and on the other hand the $Ti3d < E_F$ states play a beneficial role for the DSSC.

6.3.4 ResPES of bandgap states

Experiments at anatase single crystals conducted by Thomas et al. [131, 134] indicate, that the oxygen vacancies V_O can be resonantly excited from the Ti 3p level. Thus the V_O related states could be assigned to result from Ti 3d orbitals. Unlike the nanocrystalline anatase

6.3 Electronic structure and defects

samples used in this work, the used single crystals did not show the $Ti3d < E_F$ states. In order to gain more information about the $Ti3d < E_F$ states ResPES experiments have been performed. In this work two resonance excitations have been used, the $Ti3p \rightarrow Ti3d$ resonance (Fig.6.13 left) and the $Ti2p \rightarrow Ti3d$ resonance (Fig.6.13 right). For this purpose the excitation energy was changed in 1 eV steps across the region where the respective resonance occurs. Since the intensity of the incident photons change with time, the spectra have been normalized to the beam current.

Variation of the ionization cross section In addition to the change of photon intensity with time, the photoionization cross section of atomic orbitals varies with the excitation energy. As an approximation for the analysis the theoretical cross section of free atoms as a function of the photon excitation energy after Yeh and Lindau has been taken [274], which is shown in Fig.6.11.

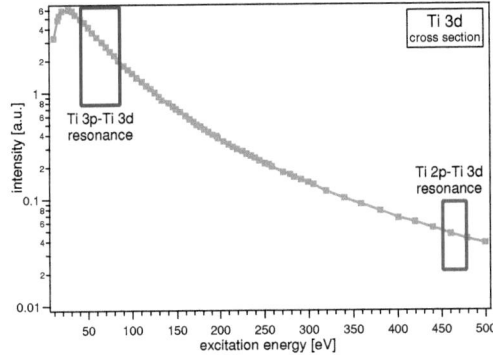

Figure 6.11: The digitized photoionization cross section data of Ti 3p as a function of the excitation energy $h\nu$ after Yeh & Lindau [274]. The frames indicate the measured area around the respective resonances.

Especially within the area of the $Ti3p \rightarrow Ti3d$ resonance the variation of the cross section is large. In order to eliminate the dependency of the photoemission signal intensity on the cross section, the ResPES spectra shown in Fig. 6.13 have been corrected by the value of the atomic cross section after being normalized to the beam current.

Additional bandgap state intensity induced by synchrotron irradiation In this work it has been observed, that additional gap state intensity was produced upon exposure to synchrotron radiation over a larger time scale. In Figure 6.12 the intensity increase of both V_O states and $Ti3d < E_F$ states as a function of exposure time to synchrotron radiation is shown. In order to minimize this dynamic effect on the ResPES measurements, the sample has been irradiated for at least 10 min with synchrotron light before taking the first spectrum. Nevertheless we observe also on not exposed samples still a certain amount of gap states, which means that the formation of additional gap states by synchrotron radiation only affects quantitative values to some extent.

Due to this dynamic creation of gap state intensity, the ResPES spectra (Fig. 6.13) have been corrected by the amount of additionally produced gap states upon exposure to synchrotron light. Since it is known, at which time each ResPES spectrum has been taken, the spectra have been divided by the relative intensity shown in Fig. 6.12 right.

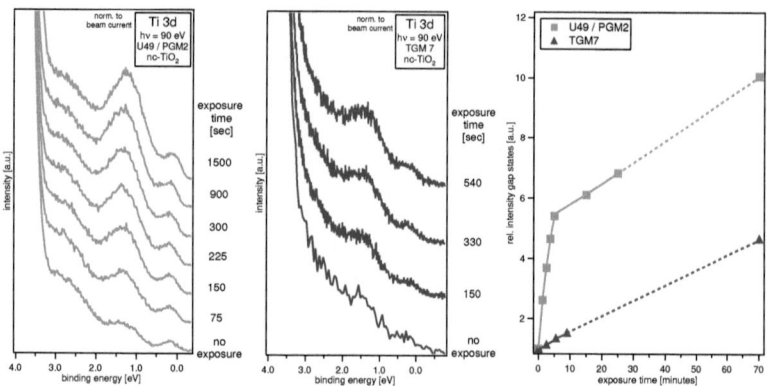

Figure 6.12: The intensity of gap states depends on the exposure time to synchrotron radiation. **left:** Photoemission spectra of the bandgap region taken at the quadrapole U49 / PGM2. **middle:** Spectra taken at the dipole TGM7. **right:** Comparison of the relative gap state intensities, starting with 1 for the not exposed sample. The values have been linearly extrapolated to 70 min exposure time.

Berger et al. [275, 276] observed by Electron Paramagnetic Resonance (EPR), that electrons, which are excited with UV irradiation to the conduction band, are trapped subsequently by Ti^{3+} states in the band gap. Since we observe an intensity correlation between the Ti^{3+} states with the $Ti\,3d\ V_O$ states in the band gap as described below, this trapping of electrons with exposure to synchrotron radiation may be applicable for the observed production of gap states.

The corrected ResPES photoemission spectra The measurements have been taken at two different beamlines: The quadrapole U49/PGM2 beamline (Fig.6.12 left), at which the photon intensity is much higher than at the dipole TGM7 beamline (Fig.6.12 middle). For the intensity correction of all spectra, the variation of photon intensity with time, the change of the ionization cross section as a function of the excitation energy and the intensity increase of gap states due to synchrotron exposure has been taken into account as described above. As it can be seen from the comparison of the relative gap state intensities, including both the V_O and the $Ti\,3d < E_F$ gap states (Fig.6.12 right), the gap state intensity increases more than 5 times during the first 5 min of exposure at the U49/PGM2 beamline, whereas at the TGM7 the intensity the gain is only about 36 %. It is noticeable, that the slope of the relative intensity growth (Fig.6.12 right) decreases with time.

Surprisingly the intensity of the $Ti\,3d < E_F$ states relative to the V_O states is much smaller in the case of the $Ti\,3p \rightarrow Ti\,3d$ resonance than for the $Ti\,2p \rightarrow Ti\,3d$ resonance. Using different photon energies for the photoemission experiments, as it is the case for the different

6.3 Electronic structure and defects

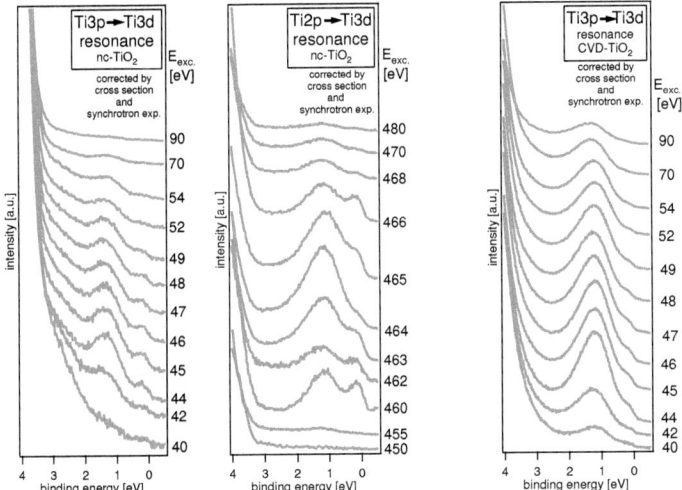

Figure 6.13: ResPES measurements showing the gap region ≤ 4 eV. **left:** The $Ti3p \rightarrow Ti3d$ resonance of nc-TiO$_2$, measured at the TGM7 beamline. **middle:** The $Ti2p \rightarrow Ti3d$ resonance of nc-TiO$_2$, which has been recorded at the U49/PGM2 beamline. **right:** For comparison the $Ti3p \rightarrow Ti3d$ resonance of CVD-TiO$_2$ is shown (taken at the TGM7 beamline).

resonance excitations, causes different inelastic mean free paths (IMFP) of electrons in the material, since the kinetic energy E_{kin} of the photoemitted electrons is changed as well. This is quantitatively displayed by the "bathtub" curve on Page 57 (Fig.3.2). Because the excitation energy for the $Ti2p \rightarrow Ti3d$ resonance is much higher than for the $Ti3p \rightarrow Ti3d$ resonance, the inelastic mean free path (IMFP) is much higher as well. For instance, at $h\nu = 50$ eV the kinetic energy of electrons from the Ti 3d orbital is $E_{kin} = 48.7$ eV. Thus the IMFP amounts to $\lambda = 5.3$ Å, whereas at $h\nu = 460$ eV ($E_{kin} = 458.7$ eV) it is 12.3 Å for anatase [269]. The surface sensitivity is much higher at smaller excitation energies, and it can be concluded, that the $Ti3d < E_F$ states are, unlike the V_O states, not located at the outermost surface. Therefore we address those states as subsurface states.

In order to quantify the ResPES spectra with respect to the resonant maximum, the intensities of the V_O and the $Ti3d < E_F$ states are displayed in Fig.6.14.

Concerning the $Ti3p \rightarrow Ti3d$ resonance, the intensity maximum for the V_O states and the $Ti3d < E_F$ states is at $h\nu = 44$ eV for both substrates. For the $Ti2p \rightarrow Ti3d$ resonance the intensity maximum for the V_O states is located at $h\nu = 464$ eV. Interestingly at the same excitation energy the $Ti3d < E_F$ state has a local minimum, which has been confirmed by repeating this particular measurement.

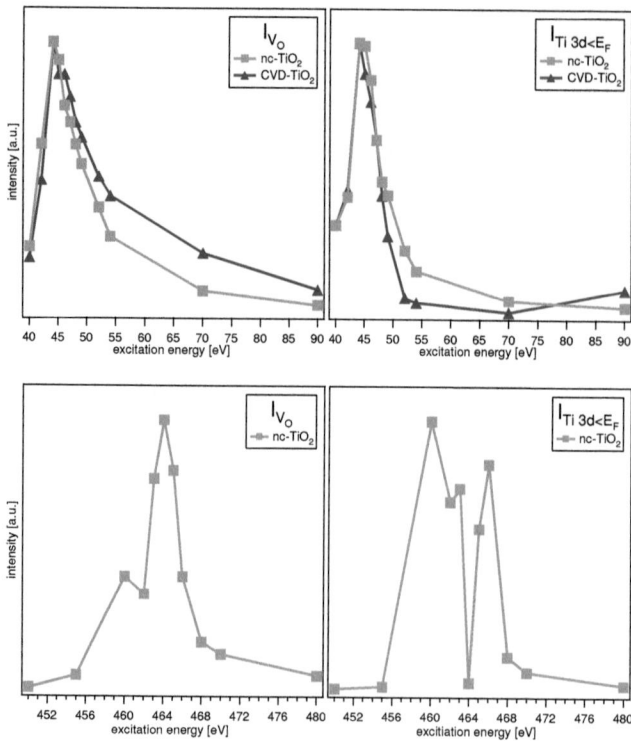

Figure 6.14: The intensity as a function of excitation energy of the spectral features for the ResPES measurements from Fig. 6.13 within the bandgap. **left top:** The relative intensity of the V_O states for the $Ti3p \to Ti3d$ resonance. **right top:** The relative intensity of the $Ti3d < E_F$ states for the $Ti3p \to Ti3d$ resonance. **left bottom:** The relative intensity of the V_O states for the $Ti2p \to Ti3d$ resonance. **right bottom:** The relative intensity of the $Ti3d < E_F$ states for the $Ti2p \to Ti3d$ resonance.

6.3.5 Discussion of the results

In the following the experimental results of this section, concerning the electronic structure and defects of the substrate, are explained in terms of three different models, shown in Figure 6.15.

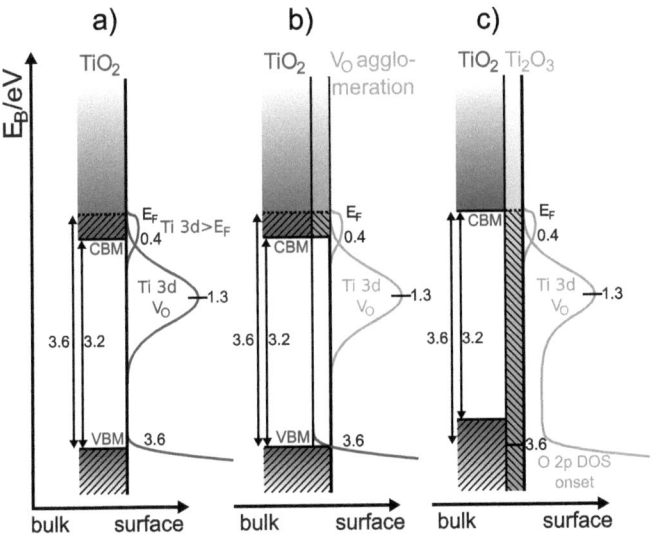

Figure 6.15: Three different models for the electronic structure of the substrate employed in this work. **a)** This model takes only one TiO_2 phase into account. **b)** The agglomeration of oxygen vacancies produce the $Ti3d < E_F$ states. This model can bee seen as an intermediate case of model a) and c). **c)** In this case the substrate material consists of two different phases.

As already mentioned in Section 6.3.3, the VBM of both anatase materials is found at $E_B = 3.6$ eV, and the oxygen vacancy states V_O are at $E_B = 1.3$ eV within the bandgap. In the case of assuming one anatase phase and the rigid band model (see Fig.6.15 a)) the Fermi level has to be 400 meV above the conduction band minimum (CBM), since the optical fundamental band gap of anatase is 3.2 eV. Thus the $Ti3d < E_F$ states are occupied states above the CBE, which implies that the semiconductor is degenerated. In addition, this model addresses the $Ti3d < E_F$ states as conduction band states. Within this model the $Ti3d < E_F$ states are regarded as delocalized and not as surface states.

A second model considers a new phase formed by defects, which coexist locally on the material, i.e. a metallic Ti_2O_3 phase on the surface and a TiO_2 bulk phase below (Fig.6.15 c)). Because of the metallic character of Ti_2O_3, the electronic states are occupied up to the Fermi level without significant bandgap. Thus at 3.6 eV we have an O 2p DOS onset instead of an VBM. Then the $Ti3d < E_F$ states originate from the conduction band of the Ti_2O_3 phase on the surface. The Fermi level is located at the CBM of the TiO_2 phase, which would mean that no occupied states are in the conduction band. Since the oxygen vacancies are addressed as surface states, they originate from the Ti_2O_3 surface phase, or they belong to the fraction of TiO_2 which is not covered by Ti_2O_3. Another hint for the existence of a Ti_2O_3 phase at the surface is given: Especially with surface sensitive measurements we observe a certain emission intensity between the O 2p onset and the oxygen vacancies, which is much lower in intensity in the spectra with lower surface sensitivity. A spectrum of Ti_2O_3, taken from literature [277], shows this DOS between the valence band onset and the gap states as well (Fig. 6.16).

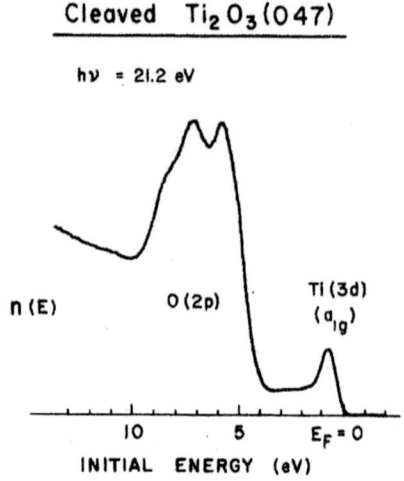

Figure 6.16: An UPS spectrum of Ti_2O_3, taken from literature [277] shows DOS between the valence band onset and the gap states.

The model shown in the middle in Figure 6.15 b) takes an agglomeration of oxygen vacancies V_O on the surface of the material into account. This model is considered being an intermediate case between the two other models described above, i.e. the material does not yet consist of two different phases. But the density of oxygen vacancies is sufficiently high to allow the formation of another "band" of electronic states resulting in the $Ti3d < E_F$ states, due to the interaction of these states with each other. Since no experimental evidence of the existence of another phase on the surface of TiO_2 is given by the performed GIXRD measurements shown in Section 6.1.1, this intermediate case is most likely representing the reality.

7 TiO$_2$/solvent interface

In this section we will have a look at the interaction of the solvent with the pristine anatase substrate. As solvents the polar solvent acetonitrile and the unpolar solvent benzene were chosen.

7.1 Adsorption of acetonitrile / benzene on TiO$_2$

Acetonitrile as the solvent for the redox couple I$^-$/I$_3^-$ is widely used, because it is an aprotic[1] and volatile solvent, which ensures a fast ion mobility. The high polarity of the solvent is responsible for a high solvation enthalpy, as shown on page 46, Table 2.2. In order to compare acetonitrile with the adsorption behavior of a solvent with low polarity, benzene has been chosen.

7.1.1 Coverage by the adsorbed solvent

Core level analysis: To estimate the coverage by an adsorbant, basically two possibilities are given in Photoelectron Spectroscopy: Either by the intensity change of characteristic emissions of the adsorbant, or by the attenuation of the substrate emissions. Because the substrate is always contamined with carbon due to its manufacturing process (Figure 7.1 right and left), the adsorbant C 1s emission line could not be taken for estimating the coverage by the adsorbed solvent. Thus the attenuation of the substrate Ti 2p signal due to the coverage by the adsorbed solvent was used for estimating the coverage assuming a flat substrate surface. In terms of this coverage estimation, Equation 3.9 (on Page 59) has been used. Since no inelastic mean free path λ (IMFP) data exists for the investigated solvents, the average value of the IMFP for organic compounds measured by Tanuma et al. [278] has been taken as an approximate value. Since the kinetic energy of electrons is around 150 eV for the Ti 2p at $h\nu = 600$ eV, the approximate IMFP of electrons in the solvent amounts to 8.3 Å.

The core level emissions of the adsorbed solvent show, that solvent molecules are actually adsorbed on the surface. In Figure 7.1 left and middle we observe an increase of intensity at $E_B = 287.3$ eV (C 1s) and 400.9 eV (N 1s), which is attributed to acetonitrile. In analogy (Figure 7.1 bottom) the increasing signal at $E_B = 285.8$ eV is due to benzene adsorption.

---

[1] In an aprotic solvent hydrogen atoms are bound covalently to carbon atoms. They cannot be split off easily, which avoids photoreactions with the sensitizing dye.

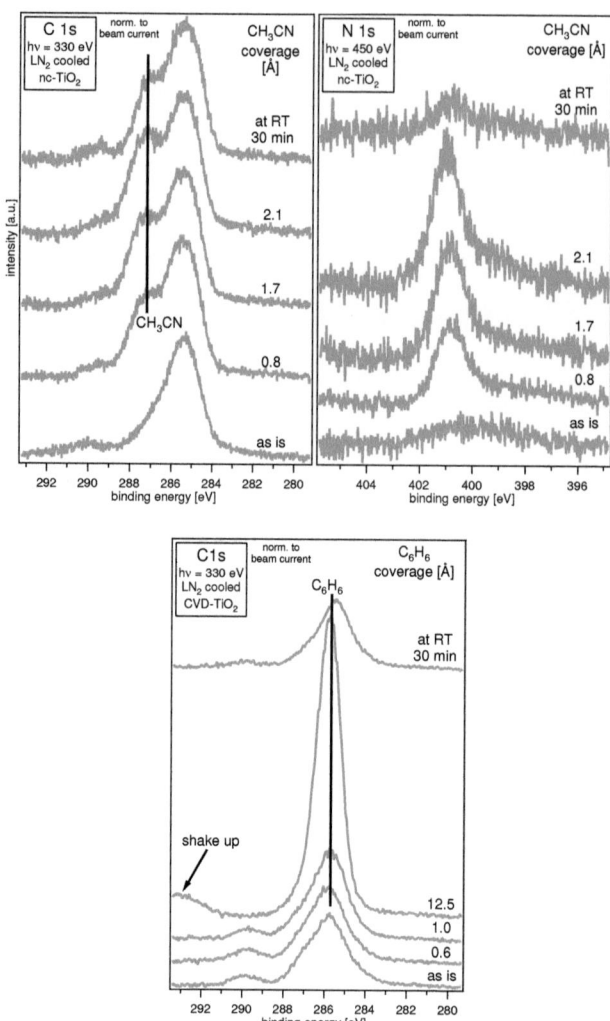

Figure 7.1: The photoemission lines of the respective adsorbed solvent measured with highest surface sensitivity ($E_{kin} \approx 50\,eV$). **top left:** C 1s core level emission of acetonitrile (CH_3CN) taken at $h\nu = 330\,eV$. **top right:** The N 1s emission line of acetonitrile ($h\nu = 450\,eV$). **bottom:** The C 1s emission line for benzene (C_6H_6) taken at $h\nu = 330\,eV$.

7.1 Adsorption of acetonitrile / benzene on TiO_2

Valence band analysis Valence band spectra were taken with an excitation energy of $h\nu = 90$ eV for both CH_3CN and C_6H_6 adsorption (Fig.7.2). The spectra were recorded from thick solvent layers on top of TiO_2 substrate and they are normalized to the Ti 3p level. Difference spectra have been calculated: The substrate spectra has been subtracted from the adsorption spectra in order to gain the pure adsorbate spectra. These solvent adsorption spectra (Fig.7.2 bottom) are compared to literature data (Fig.7.2 top). As one can see, the spectral features in the adsorbate spectra match those of the literature data.

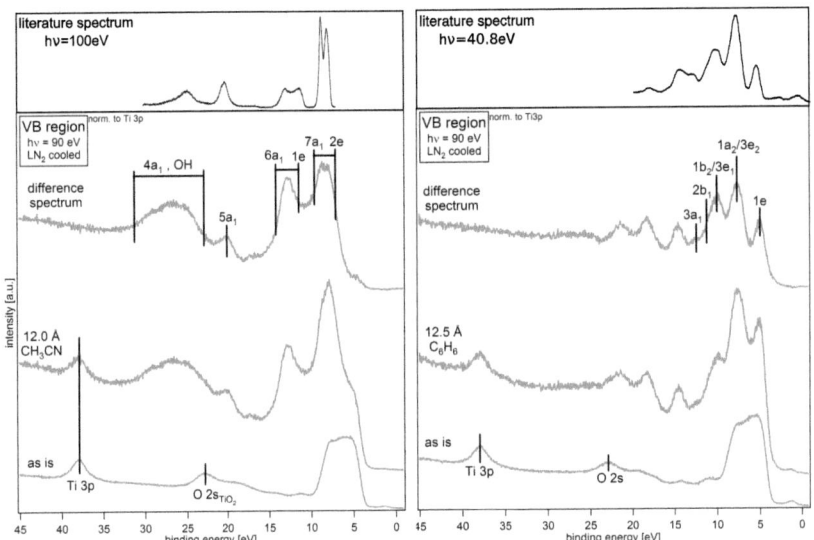

Figure 7.2: Valence band spectra of solvent adsorption on TiO_2 compared to literature data. **left:** The acetonitrile spectrum (bottom) compared to a gas phase spectrum (top) from literature (extracted from [279]). For better visibility the literature spectrum has been aligned energetically to the experimental data below. **right:** The benzene spectrum (bottom) and an literature benzene spectrum (top) from an adsorption experiment (extracted from [280]).

In the literature the binding energies of the molecular orbitals (MO) are listed for acetonitrile [279, 281] and benzene [280, 282] as obtained from photoemission spectroscopy. The spectra for actonitrile are from the gas phase, whereas the benzene literature data are from adsorption experiments. Our experimental data presented above is compared with the literature data in Table 7.1.

In the case of benzene the values in literature match very well with our data. For acetonitrile a direct comparison is not possible, since the literature values are referred against the vacuum level (ionization potential given) and our data are referred against the Fermi level. Thus the work function of the sample has to be subtracted from the literature data to obtain binding energy values. Moreover the acetonitrile spectrum does not show as well

acetonitrile			benzene		
MO	$E_{B(exp)}$ [eV]	$E_{B(lit)}$ [eV]	MO	$E_{B(exp)}$ [eV]	$E_{B(lit)}$ [eV]
$2e$	7.7 - 8.3	11.30 - 12.90	$1e$	5.1	5.1
$7a_1$	8.3 - 9.6	12.90 - 14.25	$1a_2/3e_2$	7.7	7.9
$1e$	11.4 - 12.8	14.75 - 16.85	$1b_2/3e_1$	10.0	9.8
$6a_1$	12.8 - 14.3	16.85 - 19.30	$2b_1$	11.3	11.0
$5a_1$	20.2	23.00 - 27.00	$3a_1$	12.5	12.4
$4a_1$	23.1-31.3	27.00 - 33.00			

Table 7.1: The experimental data of the binding energies of molecular orbitals of acetonitrile and benzene compared with literature values.

defined peaks as the benzene spectrum, so the values could only roughly be estimated. We observe that the values from literature are 3.7-4.8 eV at lower binding energy compared to the experimental values. On one hand the workfunction of adsorbed acetonitrile has to be taken into account. On the other hand a final state effect like screening of the photohole by electron density of the TiO_2 substrate in the case of adsorbed acetonitrile leads to a higher kinetic energy of the photoemitted electron (i.e. lower binding energy).

In the acetonitrile spectrum we suppose an OH contribution in the binding energy range of 23 - 31 eV, but it is superimposed by the $4a_1$ molecular orbital. The OH signal is probably due to water contamination. This is discussed below in the context with measured oxygen core level analysis (Section 7.1.4).

7.1.2 Interaction between solvent and oxygen defects

Ti 2p core level data Photoelectron spectra were taken of the Ti 2p emission at a photon energy of $h\nu$ = 600 eV for observing the interaction of the adsorbed solvent with the pristine substrate with high surface sensitivity (Fig.7.3). After a Shirley background substraction has been performed all spectra have been normalized to the maximum of the Ti 2p main emission peak at 459.5 eV to exclude damping due to coverage by the adsorbate.

Both acetonitrile (CH_3CN) and benzene (C_6H_6) has been adsorbed in ultrahigh vacuum (UHV), according to the experimental procedure described in Section 4.2.3. It has to be mentioned, that due to technical reasons it was not possible to adsorb benzene onto the nc-TiO_2, thus the CVD-TiO_2 has been used instead. As already presented in Section 6.3.2.1 the relative intensities of the Ti^{3+} states differ between both substrate types.

In order to quantify the intensity variation of the Ti^{3+} states relative to the Ti 2p main signal, the fitted Voigt profiles have been integrated. The integrated intensity has been normalized to the initial intensity of the untreated substrate and plotted against the estimated coverage (Fig.7.3 bottom). With adsorption of both acetonitrile (Fig.7.3 top left) as a polar solvent and benzene (Fig.7.3 top right) as an unpolar solvent we notice a decrease of intensity of the Ti^{3+} states at 1.6 eV lower binding energy compared to the main emission. Therefore

7.1 Adsorption of acetonitrile / benzene on TiO_2

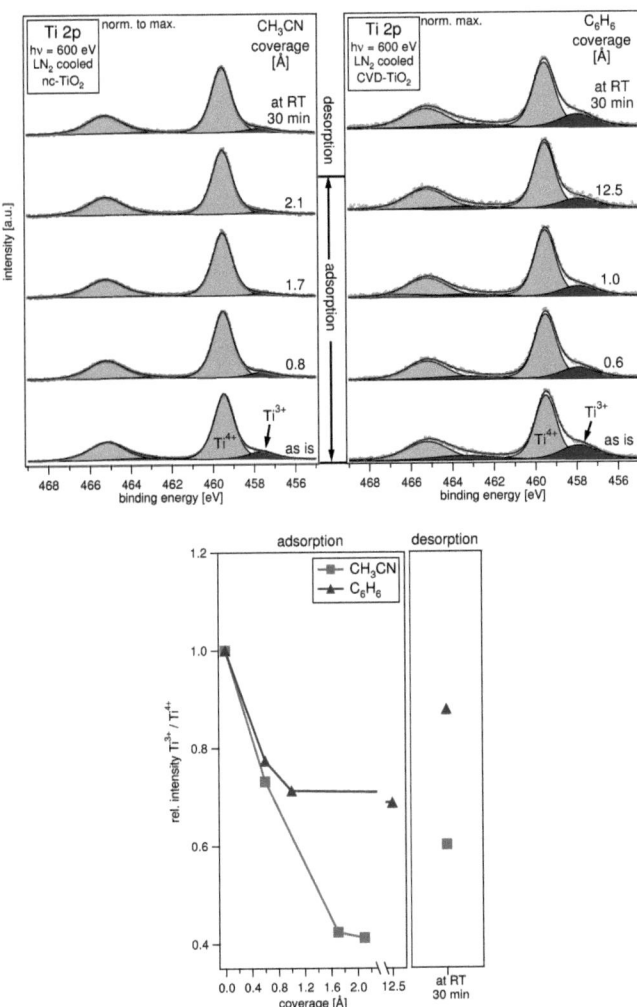

Figure 7.3: The Ti 2p core level emission line taken at $h\nu = 600$ eV **top left:** Spectra in dependence on adsorbed acetonitrile as a polar solvent. The acetonitrile (CH_3CN) coverage is indicated for each spectrum. The spectra have been fitted with Voigt profiles. **top right:** Spectra in dependence on adsorbed benzene (C_6H_6) as an unpolar solvent. **bottom:** Normalized intensities of the Ti^{3+} states in dependency on the coverage by adsorbed solvent.

it can be concluded that the solvent molecules interact with the Ti^{3+} sites on the surface. For acetonitrile adsorption the intensity decreases much faster than in the case of the benzene adsorption, which is a hint for a stronger interaction of acetonitrile molecules with the substrate compared to benzene. Due to an increase of intensity upon annealing to room temperature, a reversible of the adsorption of both solvents can be deduced. The reversibility of the adsorption indicates that both kind of molecules are not bound to the substrate by a dissociative bond formation. We suggest a donor like interaction of the lone pair electrons of the solvent molecule with the titanium atoms of the substrate.

The interaction of the Ti^{3+} sites with the solvent is in contradiction to the results of Zhuang et al. [213], who found by Infrared Spectroscopy that for the adsorption of CH_3CN onto liquid nitrogen cooled TiO_2 the vibronic frequency of the CN group matches the bonding of CH_3CN adsorbed onto Ti^{4+} sites. Thus they concluded, that no adsorption on Ti^{3+} sites occur due to sterical reasons. One can speculate about the reasons, but most likely the morphology of the not sintered TiO_2 powder used by Zhuang et al. differs from the morphology of our sintered nc-TiO_2 and CVD-TiO_2.

Ti 3d valence band features As one can see from the spectral development with adsorption (Fig. 7.4 top left and right), both solvents quench especially the oxygen vacancy states in a similar way as seen for the Ti^{3+} states in the Ti 2p emission (Fig. 7.3). Thus it is concluded, that the Ti 3d oxygen vacancy (V_O) states in the bandgap are correlated to the Ti^{3+} states.

The spectra in Figure 7.4 have been normalized to the O 2s emission at $E_B = 6.4$ eV in order to exclude damping due to coverage of the adsorbed solvent. For the Gauss-Lorentz fit a Shirley background has been considered.

In the case of acetonitrile adsorption on TiO_2 (Fig. 7.4 top left), the initial intensity of oxygen vacancy states decreases rapidly even at low coverages of 0.8 Å. Interestingly the intensity of the $Ti3d < E_F$ states stays almost at the same level, only at a coverage of 2.1 Å the $Ti3d < E_F$ states are reduced in intensity. Upon annealing at room temperature for 30 min both V_O and $Ti3d < E_F$ state gain in intensity again.

Zhuang et al. [213] have observed different adsorption sites of acetonitrile on the TiO_2 surface using IR spectroscopy. Besides the adsorption of the solvent molecules on the electron accepting Ti^{4+} sites (Lewis acid sites) as mentioned above they have observed the adsorption on proton donating OH groups on the substrate surface (Brønsted acid site bonding). The authors exclude the adsorption on Ti^{3+} states as already mentioned in Section 7.1.2. Since acetonitrile is an electron donor (Lewis base), it seems to be contradictory at the first place, that the Lewis acid site bonding on the reduced and electron donating Ti^{3+} may occur. But as we observe the quenching of the Ti^{3+} states, there must be an interaction between Ti^{3+} states and acetonitrile as well. It might be the case, that the actual reaction mechanism is not as simple as being proposed by Zhuang et al. [213] (see Section 2.3.5.1, page 50). A possible reaction mechanism for the CH_3CN adsorption onto the Ti^{3+} states is, that CH_3CN donates its nitrogen lone pair electron forming a chemical

7.1 Adsorption of acetonitrile / benzene on TiO_2

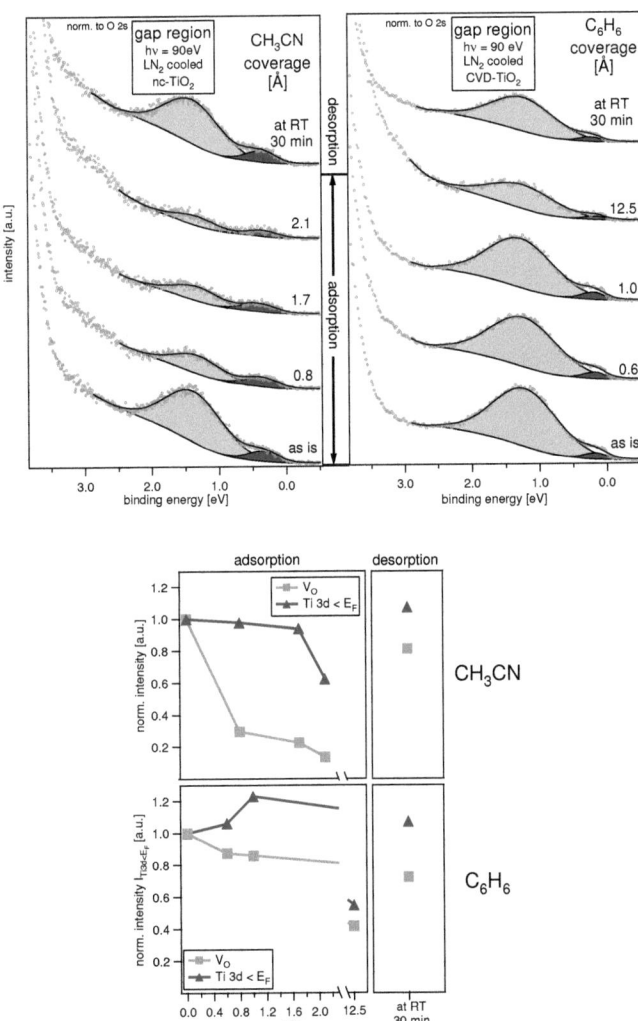

Figure 7.4: The bandgap region measured at $h\nu = 90\,eV$. **top left:** Spectra in dependence on acetonitrile adsorption. The component at higher binding energy represents the oxygen vacancy states V_O, and just below the Fermi level are the $Ti3d < E_F$ states. **top right:** Spectra in dependence on benzene adsorption. **bottom up:** The normalized intensity of the V_O and $Ti3d < E_F$ states for the acetonitrile adsorption and **bottom down:** for the benzene adsorption.

bond with the Ti sites. Consequently, the electron of the Ti^{3+} needs to be transferred into the conduction band.

In contrast to Zhuang et al, some publications propose a photochemical reaction [214, 215, 220] involving H_2O and OH on the substrate surface to form acetamide species (CH_3CONH) to form stable bonds on the substrate surface. This would imply an irreversible reaction, which contradicts the observed reversibility of the adsorption process of acetonitrile (desorption upon annealing to room temperature).

By adsorbing benzene we found a qualitative similar behaviour concerning the interaction with the gap states, but quantitatively it is completely different. The Ti^{3+} states are quenched upon adsorption of benzene as well, but we observe that much higher coverages are needed to achieve the same amount of quenching. But as one can see from the intensity dependence (Fig. 7.4 bottom), we observe, that the $Ti3d < E_F$ states just below the Fermi level are not quenched. We relate that this quantitatively different behaviour is caused by the different adsorption mechanism of the benzene molecules on the surface. For non hydroxylated rutile (110) surfaces it was observed that the benzene ring molecules lie flat on the 5-fold Ti^{4+}-sites [223, 225]. In the case of hydroxylated surfaces like on our samples the bonding to OH groups on the surface by the benzene molecules is weaker [221, 222] than the direct interaction with the Ti atoms. We also propose a donor like interaction of the ring π–system with Ti-sites, but we expect a weaker donor character compared to acetonitrile.

Accordingly to the Ti^{3+} states in the Ti 2p spectra the Ti 3d V_O states observed with photoemission on pristine TiO_2 [21, 283] are also quenched by solvent adsorption (Fig. 7.4). This parallel behavior confirms, that both oxygen vacancy states and Ti^{3+} states are correlated. Since the adsorbate is interacting with the substrate surface only, the quenching of both V_O and Ti^{3+} states asserts, that both states are located at the surface and belong to the same Ti species.

7.1.3 Work function changes

To see a change in the work function of the substrate the secondary electron edges (SEE) have been recorded at $h\nu = 50\,eV$ (Figure 7.5). The secondary electron edge is the onset of emitted electrons, which have zero kinetic energy, thus their binding energy appears at $E_B = h\nu - \phi$. Because electrons of a kinetic energy down to 0 eV may not reach the detector, a bias of 10 V has been applied between the sample and the analyzer.

It has to be mentioned, that our interest is mainly to look at trends of the work function, but not to determine its absolute values. Nevertheless, the absolute value of the work function of $\phi = 3.6 - 3.8\,eV$ matches quite well with Ultraviolet Photoemission Spectroscopy data of Orendorz et al. for pristine anatase ($\phi = 3.7\,eV$). But since the TiO_2 substrate has a nanocrystalline morphology, the surface exhibits a lot of edges and kinks with a distribution of different work functions compared to a single crystalline surface. Therefore

7.1 Adsorption of acetonitrile / benzene on TiO$_2$

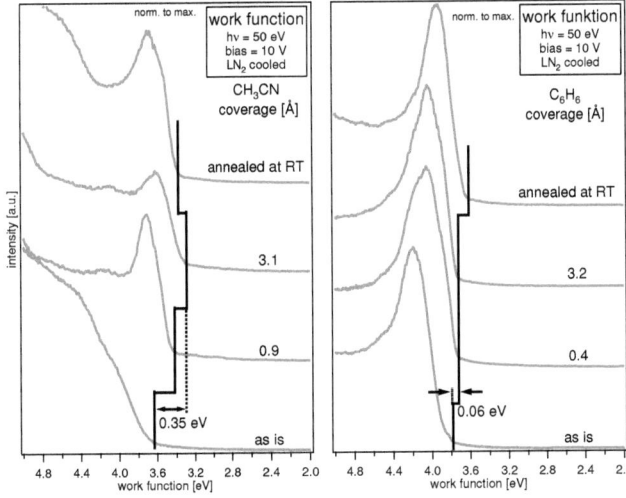

Figure 7.5: The trend of the work function for the adsorbed solvents. **left:** Acetonitrile adsorption. **right:** Benzene adsorption.

the determination of absolute values of the work function may diverge from sample to sample.

The SEE provides important hints regarding the adsorption of the acetonitrile as a polar solvent and benzene as an unpolar solvent. For analyzing the work function, the onset of the secondary edge has been taken. Inspecting the course of the SEE for both solvents, we observe a continuous shift towards lower work functions. Upon adsorption of acetonitrile it is up to 0.35 eV, whereas in the case of benzene adsorption only a very small shift of 0.06 eV occurs. Upon adsorption of more benzene the work function does not change any more. Interestingly after annealing to RT in the case of benzene adsorption the work function decreases further, whereas for the acetonitrile experiment the work function shifts back to higher values.

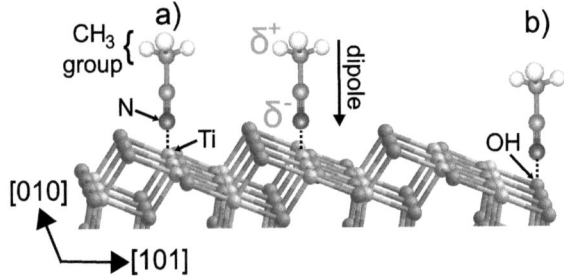

Figure 7.6: The adsorption of acetonitrile onto the untreated TiO$_2$ substrate on a **a)** Lewis acid site and **b)** Brønsted acid site. The dipole points towards the substrate surface, thus the work function is reduced.

In case of acetonitrile, the decrease of the work function upon adsorption is due to the fact, that acetonitrile is a polar molecule. Since we observed an interaction of acetonitrile with the Ti^{3+} states, we assume a Brønsted site interaction similar to Section 2.18, page 51 as depicted in Figure 7.6. The acenitrile molecules are arranged towards the sample in a way, that the positive end of the dipole points away from the substrate surface. Therefore, the work function of the substrate is reduced, because photoemitted electrons are accelerated by the dipolar drop into vacuum.

For benzene the small shift might arise from the adsorption of small amounts of molecular water onto the surface. Interestingly this trend continues upon desorption. This might be in line with the results observed at the O 1s level (Fig.7.7 right), where the OH component, which is observed for the untreated substrate, disappears upon annealing to room temperature. In other words the protonation of the surface upon annealing is lower than for the initial state of the substrate. A protonated surface, which is charged positively, leads to a downward shift of the Fermi level. Thus the deprotonation of the surface parallel to benzene desorption leads to an upward shift of the Fermi level, i.e. a decrease of the work function.

7.1.4 Solvent-oxygen interaction

O 1s core level lines   In order to investigate the solvent interaction on the oxygen site, the O 1s spectra has been recorded at $h\nu = 600$ eV excitation energy. As mentioned in Section 7.2, the untreated substrate does not exhibit molecularly adsorbed water at the surface. But at least some amount of hydroxyl is present at the surface. As one can see from Fig.7.7 top left and right, for both acetonitrile and benzene adsorptions the shoulder at higher binding energies is rising parallel to the amount of adsorbed solvent. The deconvolution into Voigt profiles shows clearly, that this is mainly due to the component being attributed to OH groups. In addition we note a very small intensity increase of a component, which may be assigned to molecularly adsorbed water. For quantification, the intensity rise of both OH and H_2O is shown in dependence of the solvent coverage in Figure 7.7 bottom. Accordingly, both OH and H_2O intensities rise linearly.

In order to attribute this intensity rise, one may think of a photoreaction of the solvent caused by the synchrotron beam. But Zhuang et al. [213] state that for a photooxidation of acetonitrile, O_2 gas is necessary, whereas the presence of OH or H_2O on the surface is not sufficient for this reaction. Since we measure the samples in ultrahigh vacuum at a base pressure of 10^{-10} mbar, no oxygen gas is present. In the case of benzene a decomposition on TiO_2 has been excluded by Zhou et al. [225] under the given conditions (LN_2 temperature, absence of an catalytic metal like palladium). Thus based on these studies, we exclude that this OH contribution is caused by a photoreaction.

If one assumes that the intensity increase of the O 1s shoulder is due to a contamination source, there are basically two possibilities, where the water could originate from: Either it comes from the ambient or the solvent is contaminated with water. Since we operate in ultrahigh vacuum at a base pressure of 10^{-10} mbar, the ambient seems to be rather unlikely to be a source for water. On the other hand ultrapure solvent has been used and the solvent source has been purified (Section 4.2.3). However, despite a very careful preparation of the adsorption it seems to be possible, that a slight contamination by water was not avoided.

7.1 Adsorption of acetonitrile / benzene on TiO_2

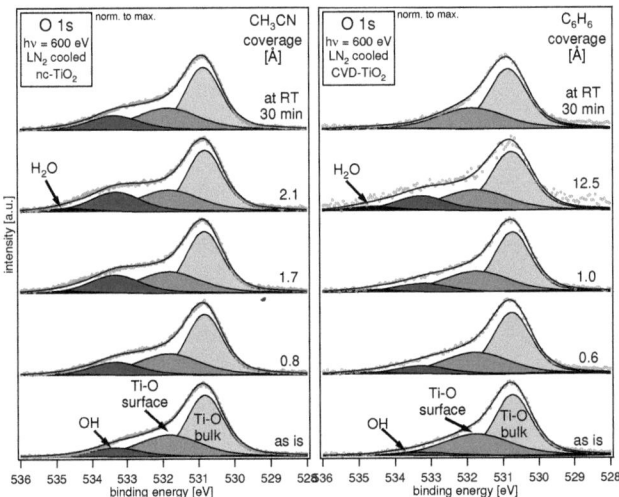

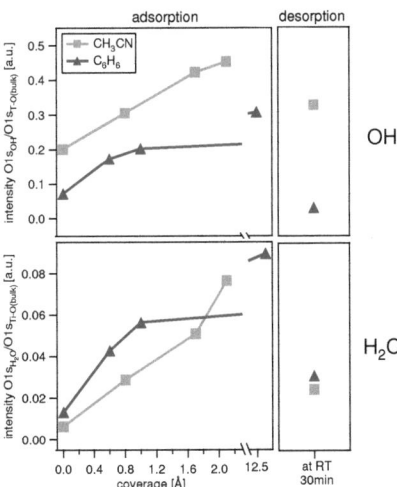

Figure 7.7: The O 1s core level spectra during solvent adsorption. **left:** The acetonotrile adsorption. **right:** The benzene adsorption. **bottom up:** The intensity of the OH- and **bottom down:** water component relative to the O 1s main signal.

Annealing to room temperature shows a reversible behavior for both solvent adsorptions. Interestingly for benzene, the desorption even goes beyond the initial state regarding the OH component. Upon desorption this component vanishes entirely.

7.2 Water adsorption

Basically water has to be considered as an unwanted side effect during the adsorption of the solvents on the sample at liquid nitrogen (LN_2) temperatures. Since at least acetonitrile is a hydrophilic solvent, the adsorption of water either out of the solvent or, which is considered being rather unlikely at a base pressure of 10^{-10} mbar, from the ambient, might occur during solvent adsorption. In order to be able to evaluate, how water induces changes in the core and valence levels and in the bandgap, adsorption experiments with water have been accomplished. In Figure 7.8 photoelectron spectra for the adsorption of water on nc-TiO_2 are displayed.

Looking at the O 1s core level (Fig. 7.8 top) shows molecularly adsorbed water (a thick layer) at a binding energy of 534.6 eV. Thus molecularly adsorbed water can be excluded being initially present on the substrates as shown in Section 6.9.

The stepwise adsorption of water is evidenced by the valence band spectra, shown in Figure 7.8 bottom left. At higher coverages extra emissions belonging to well known three molecular orbitals of molecularly adsorbed water can be clearly identified: The nonboning $1b_1$ state at $E_B \approx 9$ eV, the partly bonding $3a_1$ state at around 11.5 eV and the bonding $1b_2$ state at around 15 eV binding energy. The involved atomic orbitals are listed by Thiel et al. [284]. These emission lines have been observed in several photoemission studies [187, 285–289]. The Ti 3p and the O 2s emission belonging to TiO_2 is damped, whereas the O 2s level belonging to H_2O (the $2a_1$ orbital) increases in intensity. As one can see the adsorbate emissions shift towards higher binding energy, whereas the substrate emissions do not shift. Mayer [187] has found, that this adsorbate shift is caused by the creation of an interface dipole between the substrate and the adsorbed water molecules. Furthermore Mayer found, that the ionization potential of the adsorbed water is constant.

Of special interest for this work the bandgap spectra are shown by Figure 7.8 bottom right. They have been recorded at $h\nu = 50$ eV close to the $Ti3p \rightarrow Ti3d$ resonance and normalized to the beam current. The normalization to the Ti 3p level was not possible, since with 50 eV excitation energy the Ti 3p level cannot be excited. Alternatively the spectra have been normalized to the beam current. It has to be noted, that the beam current normalization does not exclude the damping by the adsorbate. Therefore the intensity of the bandgap states is underrated by the amount of damping. The absolute intensity of the V_O states is decreasing by adsorption of water, as it can be seen at lower coverages. Interestingly they seem to increase again, beginning at the adsorption step at a coverage of 1.8 Å, which are exceeding the original intensity at the 8.3 Å coverage by far.

But the intensity rise of the states in the bandgap is due to a different kind of electronic

7.2 Water adsorption

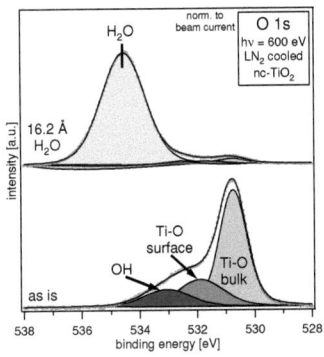
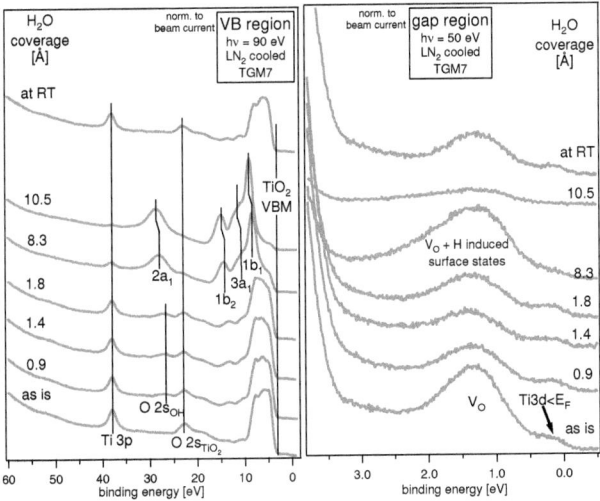

Figure 7.8: Core and valence level spectra of water adsorbed on nc-TiO_2. For details see text. **top:** O 1s core level spectra of a thick adsorbed water layer recorded at $h\nu = 600$ eV. **bottom left:** The valence band spectra with a stepwise adsorption of water ($h\nu = 90$ eV). **bottom right:** The bandgap spectra have been taken at $h\nu = 50$ eV close to the $Ti3p \rightarrow Ti3d$ resonance. It has been normalized to the beam current.

states caused by the adsorption of water. They have been addressed being hydrogen induced surface states by Feibelman et al [285] in a photoelectron spectroscopy study on titanium (0001). In that study with further adsorption these hydrogen induced surface states decrease, which is confirmed by our results. A similar intensity rise of bandgap states was also found on rutile surfaces by Kurtz et al. [290] and Di Valentin et al. [291]. For us the interesting point from that is, that an unwanted coadsorption of water coming along with the adsorption of solvent layer would lead to an apparent increase of bandgap state intensity. But as we could see from Figure 7.4, the bandgap state intensity is decreased by the adsorbed solvents (CH_3CN or C_6H_6).

7.3 Discussion of the results

In the bandgap spectra, we observe a large decrease of V_O states by adsorbing acetonitrile, whereas the intensity of the $Ti3d < E_F$ states is only slightly affected (Figure 7.9). Therefore the adsorbed acetonitrile evidently has a different influence on the two kinds of bandgap states. The slope of the intensity decrease of the Ti^{3+} states lies in between the two bandgap state intensity dependencies. This suggests, that the Ti^{3+} states are correlated with both V_O and $Ti3d < E_F$ states.

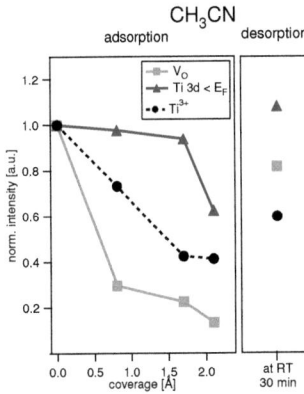

Figure 7.9: The intensity dependence of the V_O, $Ti3d < E_F$ and Ti^{3+} states of the acetonitrile adsorption onto the bare TiO_2 substrate.

Surface states are commonly addressed [48, 76, 78, 144, 145] to play a deleterious role for the DSSC in a way, that they provide a recombination path for photoinjected electrons (Fig.7.10).

Since the DOS of the V_O states is quenched strongly and the $Ti3d < E_F$ states are only weakly quenched by acetonitrile, this solvent has a beneficial impact on the DSSC. By the strong quenching of the V_O states, the adsorbate prohibits to some degree electrons to recombine with the holes in the HOMO level of the sensitizing dye. On the other hand the weak passivation by adsorbed acetonitrile with the $Ti3d < E_F$ states allows the conclusion, that these $Ti3d < E_F$ states are not located at the surface. Probably these states increase the conduction of the photoinjected electrons to the cell contact, being beneficial for the cell performance.

The performed work function measurements (Fig. 7.5) evidence, that adsorbed acetonitrile forms a dipole pointing towards the substrate (Fig. 7.6), which means, that acetonitrile interacts with its nitrogen atom with coordinatively nonsaturated Ti^{3+} states.

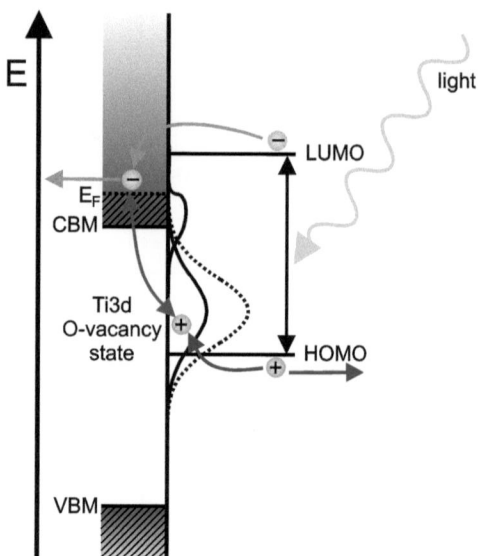

Figure 7.10: Recombination of photoinjected electrons via oxygen vacancy states. Upon solvent adsorption the oxygen vacancy state intensity (dashed line) is quenched to some extent (solid line).

8 TiO$_2$/N3 Ru dye interface

In this chapter the sensitizing dye RuL'$_2$(NCS)$_2$ is characterized with respect to its chemical components, which are resolved in the photoelectron spectra. Appropriate core level spectra have been taken with highest surface sensitivity. Also the interaction of the dye with the TiO$_2$ substrate has been investigated.

8.1 Components of the dye

The adsorption of the sensitizing dye on TiO$_2$ has been accomplished as described in Section 4.2.2. By taking spectra of the appropriate core level emissions belonging to the dye atoms, the components of the dye are quantified in the following.

8.1.1 Carbon and Ruthenium

The components of the spectra In order to identify the different components of the dye, the C 1s signal has been measured. For recording the C 1s spectra an excitation energy of $h\nu = 330$ eV yielding highest surface sensitivity has been chosen, since we obtain only monolayer coverage of the dye on the substrate.

As a substrate insitu prepared CVD-TiO$_2$ has been chosen in order to avoid contamination. But as one can see there is a small contamination level present on the untreated substrate, which originates from the sample preparation process by itself. It is a reaction product and cannot be removed completely by heating. To obtain the "pure" C 1s spectra of the dye sensitized sample (Fig.8.1 left, top), the substrate spectrum has been subtracted from the dye spectrum. The spectra have been normalized to the Ti 3d emission to exclude attenuation of the signal by the carbon contamination.

The C 1s spectra shows several chemically shifted components. In order to assign the different components within the C 1s signal, a fit by Voigt profiles has been performed. The fit consists of three C 1s emission line profiles and one doublet displaying the contribution from the Ru 3d emission. The spin orbit split between the Ru 3d$_{5/2}$ and the Ru 3d$_{3/2}$ level amounts 4.2 eV [239]. Because the Ru 3d$_{3/2}$ signal is superimposed completely by the C 1s signal, its intensity has been assumed by the intensity ratio of 2:3 for d-orbitals.

Similar to Liu et al. [292] the binding energy of the Ru 3d$_{5/2}$ signal is $E_B = 281.6$ eV. Compared to the results of Johansson et al. [159], the binding energies of the Ru 3d and the C 1s signal are shifted about 0.6 eV towards higher values. This binding energy difference

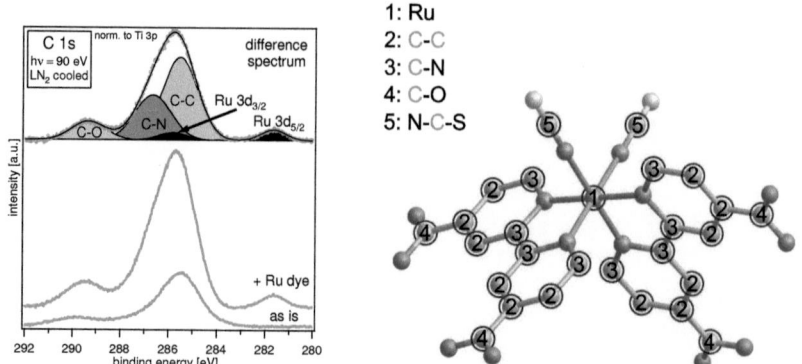

Figure 8.1: left: The C 1s emission ($h\nu = 330\,\text{eV}$) of the dye adsorbed on the TiO$_2$ substrate. **right:** The molecular structure of the N3 dye showing the positions of the different carbon atoms (marked atoms).

of 0.6 eV arises from different binding energy normalization: Johansson et al. aligned the binding energy by setting the Ti 2p$_{3/2}$ to 458.6 eV, whereas we use the Fermi edge for the binding energy alignment. The different carbon components have been addressed according to Liu et al. [292]: Aromatic carbon (C–C) at $E_B = 285.5\,\text{eV}$, C–N carbon at 286.7 eV and carboxyl carbon (C–O) at 289.4 eV. The carbon of the NCS (thiocyanate) group (N–C–S) has not been fitted separately, since it is located very close to the C–N component. Therefore it is not feasible to separate it from the C–N component.

Relative atomic concentrations In the following the atomic concentration of the respective carbon atoms in the dye molecule relative to the ruthenium center atom has been estimated. By means of the Voigt profiles of the respective emission lines the relative concentrations of the atoms in the molecule can be determined according to Equation 3.8 (page 59). One has to consider, that the atomic sensitivity factors (ASF) [242] have been determined for solid elements, an excitation energy of Al K$_\alpha$ $h\nu = 1486.6\,\text{eV}$ and an hemispherical analyzer (*Phoibos 150 MCD-9* from *Specs*). But like the cross section the ASF changes by variation of the kinetic energy (Equation 3.7). Therefore the ASF has to be determined for both the C 1s and Ru 3d emission at $h\nu = 330\,\text{eV}$.

Both C 1s and Ru 3d signals are close to each other with respect to binding energy, so that the kinetic energy E_{kin} and the inelastic mean free path (IMFP) of electrons λ_e is approximately the same. Thus the ASF for both Ru 3d and the C 1s signal at $h\nu = 330\,\text{eV}$ compared to the ASF at 1486.6 eV changes only by the cross section ratio

$$ASF(X)_{330eV} = ASF(X)_{1486.6eV} \frac{\sigma(X)_{330eV}}{\sigma(X)_{1486.6eV}} \qquad (8.1)$$

8.1 Components of the dye

where X stands either for the C 1s or Ru 3d emission line. In Figure 8.2, the cross sections as a function of the excitation energy are given for both C 1s and Ru 3d emission line.

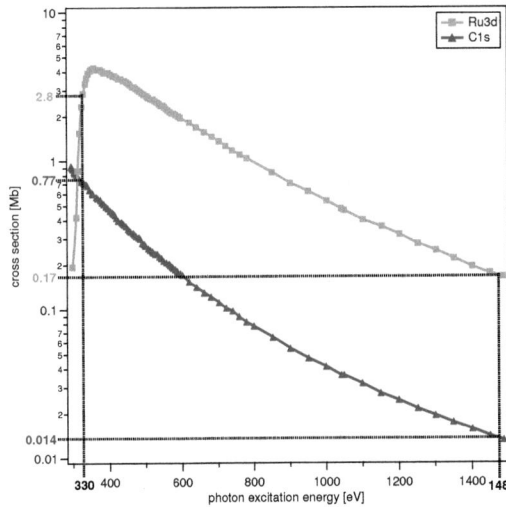

Figure 8.2: The cross section of both C 1s and Ru 3d states as a function of the photon excitation energy $h\nu$ [239]. For better visibility the respective cross sections for the appropriate excitation energies are marked.

	C 1s$_{1486.6eV}$	C 1s$_{330eV}$	Ru 3d$_{1486.6eV}$	Ru 3d$_{330eV}$
σ_A [Mb]	0.014	0.77	0.17	2.8
ASF	1.000	55.000	12.547	206.649

Table 8.1: The ASF at different excitation energies, which is normalized versus the C 1s emission at $h\nu = 1486.6$ eV [242].

By using the ASF of the respective emission line at the excitation energy of $h\nu = 330$ eV (see Table 8.1) we can calculate the relative concentrations following Equation 3.8:

$$\frac{n_{Ru}}{n_C} = \frac{I_{Ru3d330eV} \cdot ASF_{C1s330eV}}{I_{C1s330eV} \cdot ASF_{Ru3d330eV}} \tag{8.2}$$

In that way the obtained intensity corresponds to the number of atoms in the dye molecule. Since we have one Ru atom in the dye molecule, we get the experimentally determined number of atoms in the dye molecule $N_{exp.}$.

	C–C	C–N	C–O	Ru 3d
E_B [eV]	285.5	286.7	289.4	281.6
N_{theory}	12	10	4	1
$N_{exp.}$	22.5	12.6	4.9	1

Table 8.2: The experimentally determined number of carbon atoms in the dye molecule $N_{exp.}$ compared to the number of atoms from the dye molecular structure N_{theory}. The number of N–C–S atoms is not listed separately, but it is included in the number of C–N atoms.

Except the C–C component, the experimental data matches reasonably to the number of atoms from the dye molecular structure N_{theory} (compare with Fig.2.10, page 40). The strong deviation of the C–C component can be explained by carbon contamination, since the dye was adsorbed from ethanol solution. Since the dye cannot be deposited by a clean vapor deposition process, this contamination is not avoidable.

8.1.2 Nitrogen

The components of the spectra The central atom of the N3 dye is octahedrally coordinated by six nitrogen atoms. In order to characterize the different dye components it is of interest to measure the N 1s signal. In order to yield highest possible surface sensitivity an excitation energy of $h\nu = 450$ eV has been chosen.

The N 1s photoemission signal is shown in Figure 8.3. Unlike in the case of the C 1s emission line no contamination of nitrogen comes along with the substrate. The N 1s signal consists of at least two components, one about twice as big as the other. The larger emission at higher binding energy ($E_B = 400.8$ eV) is attributed to the nitrogen atoms being located in the bipyridine ligand. The smaller emission at $E_B = 398.7$ eV is due to the nitrogen atoms from the thiocyanate group. In accordance with Johansson et al. [159] and Liu et al. [292] both emissions are split by 2.1 eV. The absolute binding energies match with the ones published by Liu et al. [292], but again differ by about 0.7 eV towards higher binding energies from Johansson et al. [159]. This binding energy difference of 0.7 eV arises again from the different binding energy normalization described above (Section 8.1.1). Apparently both emission peaks are slightly asymmetric, which is due to an additional component.

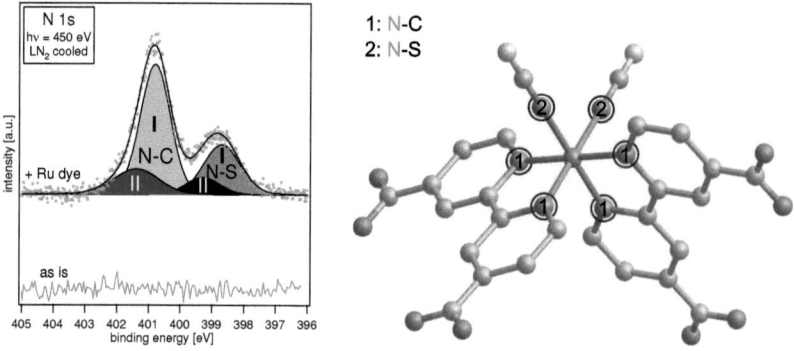

Figure 8.3: left: The N 1s emission line taken at $h\nu = 450$ eV. I and II are attributed to non interacting and interacting dye molecules (see text). **right:** The positions of the N–C and N–S nitrogen atoms within the molecule.

In the Voigt fit (Fig.8.3) a smaller pair of components (marked with II) is 0.6 eV shifted towards higher binding energy relative to a larger pair, which is due to different chemical environments around the dye molecules. We assign the component I to the fraction of dye molecules, which are not interacting with each other. This is discussed in detail in the context of the solvent coadsorption in Chapter 9. In contrast the small pair is addressed to

8.1 Components of the dye

interacting dye molecules and to molecules, which interact with the TiO_2 substrate surface. The ratio of both pairs with respect to their integrated intensity amounts about 1:3.

Relative atomic concentrations As before for the C 1s emission we now compare the intensities of the N 1s line with the Ru 3d integrated intensity. In order to do that, we need to know the ASF for N 1s at $h\nu = 450$ eV. Since we have for the N 1s emission line at 450 eV the same E_{kin} as for the Ru 3d emission at 330 eV, the ASF change only – as in the case for C 1s – by the cross section. Thus we can calculate the ASF for N 1s analogous to Equation 8.1:

$$ASF_{450eV,N1s} = ASF_{1486.6eV,N1s} \frac{\sigma_{450eV,N1s}}{\sigma_{1486.6eV,N1s}} \quad (8.3)$$

The cross sections for N 1s and the respective ASF are shown in Table 8.3.

	N $1s_{1486.6eV}$	N $1s_{450eV}$
σ_A [Mb]	0.025	0.57
ASF	2.100	47.882

Table 8.3: The ASF at different excitation energies.

In order to determine the relative intensities of the N 1s components relative to the Ru 3d emission, both Ru 3d and N 1s emission lines have to be normalized to the beam current. In addition the intensities of the spectra need to be corrected by the photon flux of the U49/PGM-2 beamline, which is about 30 % higher for the photon energy of $h\nu = 450$ eV (N 1s) compared to 330 eV (C 1s), as one can see from Figure 8.4.

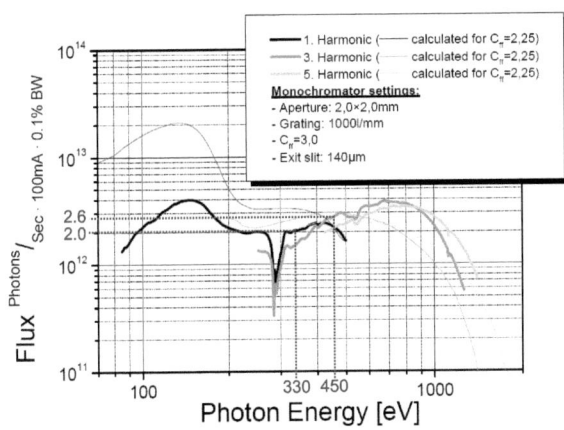

Figure 8.4: The photon flux at the U49/PGM-2 beamline as a function of the photon excitation energy. For better visibility the values for the photon energy of 330 eV and 450 eV have been marked.

The obtained relative atomic concentrations compared to the ruthenium concentration are displayed in Table 8.4.

As one can see from the table, the experimental relative atomic concentrations match reasonably and deviate by only about 10 % from the theoretical values. In addition the ex-

CHAPTER 8. TIO₂/N3 RU DYE INTERFACE

Table 8.4: The experimentally determined number of carbon atoms in the dye molecule $N_{exp.}$ compared to the number of atoms from the dye molecular structure N_{theory}.

	N–C	N–S	Ru 3d
E_B [eV]	400.8	398.7	281.6
N_{theory}	4	2	1
$N_{exp.}$	3.7	1.8	1

perimentally obtained ratio of N–C vs. N–S component match very well with the theoretical ratio of 2:1.

8.1.3 Oxygen

The components of the spectra Oxygen atoms are located in the carboxyl (COOH) groups of the dye. According to literature preferably two of the four carboxyl groups are attached onto the TiO₂ substrate under deprotonation [165–167]. Therefore two oxygen atoms are directly attached to the substrate. The other two carboxyl groups are presumably protonated and not bound to the substrate surface. The O 1s spectra have been taken at $h\nu = 600$ eV and were normalized to the beam current (Fig.8.5). Upon dye adsorption one can observe a strong decrease in intensity of the O 1s emission line belonging to the underneath TiO₂ substrate. In addition the shoulder at higher binding energy increases in intensity.

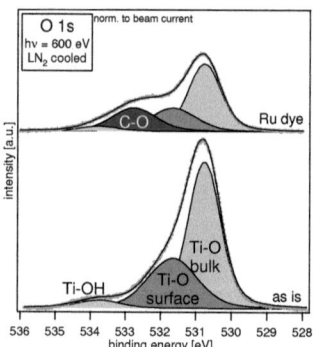
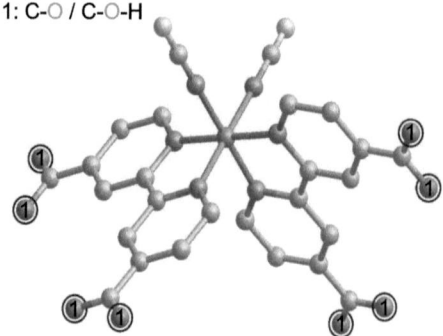

Figure 8.5: **left:** The O 1s emission of the bare substrate and dye taken at $h\nu = 600$ eV. **right:** The molecular structure of the N3 dye showing the positions of the oxygen atoms (marked atoms).

By applying a Voigt fit to the spectra the different oxygen components of the TiO₂ substrate are shown (according to Section 6.3.2.2, page 87). With dye adsorption an additional component at $E_B = 532.8$ eV appears, which is attributed to the carboxyl group of the outermost parts of the dye. In the O 1s spectra we can not distinguish between both protonated and unprotonated oxygen species. Compared to Johansson et al [159], it lies 1.3 eV at higher binding energies, whereas the Ti–O bulk peak is shifted only 0.8 eV. But in the fitting procedure made by Johansson et al, no bridging oxygen component was considered, hence the

8.1 Components of the dye

therein attributed carboxyl profile has to cover the intensity, which is close to the Ti–O bulk peak. This leads to the shift to lower binding energies.

Relative atomic concentrations Like for the quantification procedures presented before, the relative atomic concentrations have been calculated by means of the O 1s core level emission line. Since the kinetic energy of electrons for the O 1s line taken at $h\nu = 600\,\text{eV}$ ($E_{kin} = 69\,\text{eV}$) differs from the Ru 3d emission taken at $h\nu = 330\,\text{eV}$ ($E_{kin} = 48\,\text{eV}$), the transmission function of the analyzer $T \propto \frac{1}{\sqrt{E_{kin}}}$ has to be taken into account. Also the inelastic mean free path of electrons λ_e is needed. Thus the ASF has to be calculated according to Equation 3.7:

$$ASF_{600eV,O1s} = ASF_{1486.6eV,O1s} \frac{\left[\sigma \lambda_e \frac{1}{\sqrt{E_{kin}}}\right]_{600,O1s}}{\left[\sigma \lambda_e \frac{1}{\sqrt{E_{kin}}}\right]_{1486.6,O1s}} \quad (8.4)$$

The ASF with all needed parameters are listed in Table 8.5.

	O 1s$_{1486.6eV}$	O 1s$_{600eV}$
σ_A [Mb]	0.039	0.41
E_{kin} [eV]	956	69
λ_e [Å]	29.4	6.9
ASF	4.211	38.758

Table 8.5: The ASF of O 1s at different excitation energies and all parameters, which were needed to calculate the ASF.

By using Equation 3.8 (page 59) the relative atomic concentrations are obtained. It is important to mention that the intensities have to be corrected by the photon flux, which is 70 % higher at 600 eV compared to $h\nu = 330\,\text{eV}$. The obtained atomic concentrations are listed in Table 8.6.

	O–C	Ru 3d
E_B [eV]	532.8	281.6
N_{theory}	8	1
$N_{exp.}$	10.6	1

Table 8.6: The experimentally determined number of oxygen atoms in the dye molecule $N_{exp.}$ compared to the number of atoms from the dye molecular structure N_{theory}.

The experimentally calculated relative concentration of carboxyclic oxygen exceeds the theoretical concentration by more than 30 %. Since the dye is adsorbed from ethanol solution onto the TiO$_2$ substrate, this large deviation arises most likely from unwanted contamination by the ethanol solvent. The comparison with the O 1s spectra published by Johansson et al. [159] reveals, that other groups have this contamination at an even larger scale.

8.1.4 Sulphur

The S 2p orbital is appropriate to learn more about the dye-substrate interaction, because it is located at the outermost position of the thiocyanate group on the dye molecule. The

HOMO of the dye is attributed to the SCN group in several publications [9, 151, 152, 293]. Since no other sulfur atoms are present in the dye molecule, its emission line can be well distinguished from other parts of the dye molecule. Analogous to the C 1s and N 1s signal the excitation energy ($h\nu = 210$ eV) has been chosen to yield maximum surface sensitivity. The S 2p emission of the dye is a broad emission line, which we deconvoluted into three components, located at $E_B = 162.5$ eV (I), 163.3 eV (II), and 164.2 eV (III), respectively.

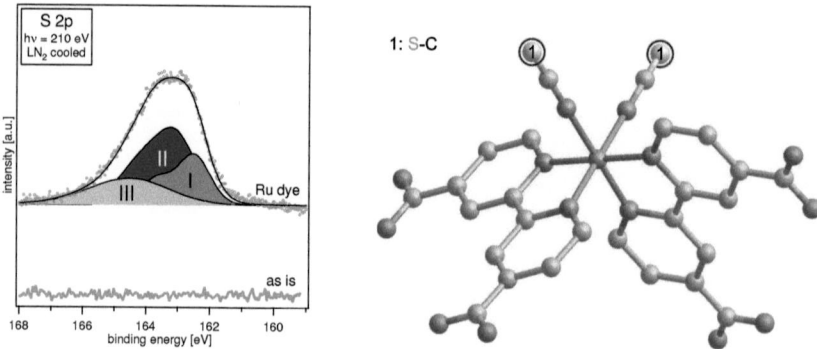

Figure 8.6: left: The S 2p emission line measured at $h\nu = 210$ eV with voigt profiles. **right:** The sulfur atom positions in the N3 dye molecule.

While the binding energy difference between component I and II is very close to the results of Johansson et al. [159] the absolute binding energies are shifted 0.6 eV to higher binding energies like the other core level emissions presented above. The components I and II are assigned to non-interacting and interacting sulphur atoms, respectively, by Johansson et al. Since Johansson et al. assign the non-interacting component at the same binding energy as the S 2p emission of the multilayer, they do not account for an interaction of the sulphur atoms with other dye molecules. According to to Johansson et al. [159] we attribute component I to the NCS groups pointing away from the surface (page 42, Fig.2.12 b) - d)) and component II to the fraction of NCS groups interacting with the surface of the anatase substrate (Figure 2.12 e) and f)). In line with Johansson et al. we suggest, that the interaction of the electronegative sulphur atom with electropositive titanium atoms of the TiO_2 substrate leads to a shift towards higher binding energies. The assignment of the third component III is somewhat unclear, but tentatively we address it to sulphur atoms lost from thiocyanate groups due to exposure to synchrotron light. Also from the S 2p spectra published by Johansson et al. [159], this third component is visible, but has not been considered in his publication.

Table 8.7: The binding energies and the fraction of the components of the S 2p emission line.

	I	II	III
E_B [eV]	162.5	163.3	164.2
fraction [%]	27	49	24

The dye orientation is crucial for the performance of the Dye Sensitized Solar Cell, as described already on page 41 (Section 2.2.2). To ensure an effective charge separation,

8.2 Interaction with the substrate

the dye molecule needs to be orientated in a way, that the HOMO level, located on the SCN group, points away from the substrate. But as we see from Table 8.7, only 27 % of the molecules are orientated in the favorable way. About half of the dye molecules interact with each other or with the substrate, and the tentatively assigned decomposed sulphur of dye molecules amounts to 24 %.

Relative atomic concentrations In order to quantify the element concentration for sulphur relative to ruthenium, again the ASF for sulfur needs to be calculated. Because the kinetic energy for the S 2p at $h\nu = 210$ eV and the Ru 3d at 330 eV is about the same, only the cross sections for calculating the ASF are needed (Table 8.8).

	S $2p_{1486.6eV}$	S $2p_{210eV}$
σ_A [Mb]	0.023	3.8
ASF	1.463	241.713

Table 8.8: The ASF of S 2p at different excitation energies.

According to Figure 8.4, the photon flux for $h\nu = 210$ eV is about 5 % higher than for 330 eV. Applying this slight correction leads to the results shown in Table 8.9.

	S-C	Ru 3d
E_B [eV]	162.5 - 164.2	281.6
N_{theory}	2	1
$N_{exp.}$	2.2	1

Table 8.9: The experimentally determined number of sulphur atoms in the dye molecule $N_{exp.}$ compared to the number of atoms from the dye molecular structure N_{theory}.

As one can see, the sulphur to ruthenium ratio fits reasonably with the theoretically predicted value. The experimentally determined number exceeds the theoretical value by 10 %, which may be due to the fact, that the sulphur atoms are the outermost atoms of the molecule pointing towards vacuum.

8.2 Interaction with the substrate

8.2.1 HOMO and oxygen vacancies

Since the dye anchors with its carboxyclic groups onto the TiO$_2$ substrate surface [165–167], the interaction between dye and TiO$_2$ must be observable. By measuring the gap region ($h\nu = 90$ eV) one can observe the behavior of the oxygen vacancies of TiO$_2$ upon dye adsorption. Of course the HOMO level of the dye is of special interest. As shown by electronic structure calculations performed by Rensmo et al [150] (page 40, Figure 2.10), the HOMO is mainly located on the thiocyanate group of the dye. The gap region spectra have been normalized to the Ti 3p emission line in order to eliminate attenuation by the adsorbed dye.

The spectrum of the untreated TiO$_2$ (Fig. 8.7 left) shows the oxygen vacancies V_O at $E_B = 1.3$ eV and the states just below the Fermi level $Ti3d < E_F$. Upon dye adsorption

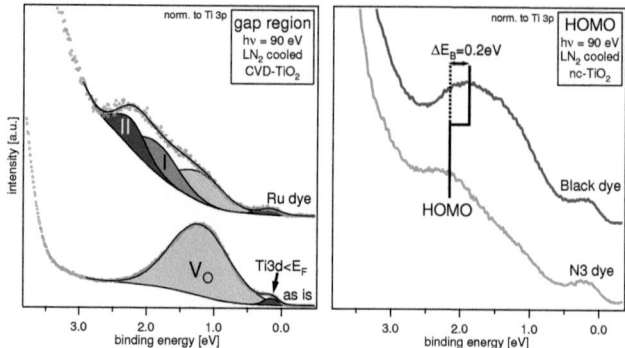

Figure 8.7: Band gap spectra after Voigt fits have been applied to the gap region. **left:** The gap region measured at $h\nu = 90$ eV. **right:** The gap region measured at $h\nu = 90$ eV, which shows the HOMO level of two different dyes (N3 dye and black dye)

additional emission lines appear at higher binding energy at around $E_B = 2$ eV, which are attributed to the dye HOMO level. Liu et al. found the HOMO level at $E_B = 1.8$ eV. This discrepancy arises from a different position of the valence band maximum. While Liu et al. observed it at $E_B \sim 3.3$ eV, we find it at $E_B = 3.6$ eV.

Applying a Voigt fitting procedure for the untreated substrate the oxygen vacancies V_O at 1.3 eV and the states just below the Fermi level $Ti3d < E_F$ have been assigned as in the chapters before. Very clearly one can see the decrease of intensity of the oxygen vacancies V_O of the TiO_2 substrate upon dye adsorption, which shows an analogous behavior as upon solvent adsorption (Section 7.1.2). The intensity of the V_O states with adsorbed dye amounts to 32 % of their original intensity, while the $Ti3d < E_F$ states stay almost the same in intensity.

Regarding the dye it has been assumed, that the HOMO level is split into two components. These components are separated by 0.4 eV, which is similar to the split of 0.6 eV of the components I and II of the S 2p level (Fig.8.6). The components I and II of the HOMO level are assumed to be of the same origin as the respective S 2p components: Within this assumption, the component I is from the isolated molecules, not interacting with each other, whereas the second component II at higher binding energy arises from the interaction of dye molecules with each other or the substrate. The reason for the 200 meV smaller split for the HOMO level in comparison to the split assigned at the S 2p level might be due to the fact that the distribution of the HOMO valence level reaches over the thiocyanate group and the metal central atom. That means that the HOMO valence level is not as strongly influenced by an interatomic interaction like the outermost sulphur atom.

It has to be mentioned, that the background subtraction for the spectrum of the adsorbed dye has to be scrutinized, because the slope of the subtraction is too high due to additional induced states at the VBM. Nevertheless the statement of a relative decrease of the V_O states is still correct, even without background subtraction.

8.2 Interaction with the substrate

For comparison the HOMO of another sensitizing dye is shown in Figure 8.7 right. This dye is $Ru^{II}(4,4',4''\text{-tricarboxy-2,2':6',2''-terpyridine})(NCS)_3$, shortly called "black dye" because of its dark color. This dye absorbs light down to the IR region at 920 nm, whereas the N3 dye absorbs only down to 750 nm. The HOMO level is located at about 200 meV lower E_B compared to the N3 dye. Since the N3 dye is much more widely used for the DSSC, it has been used in this work. Moreover the fact that the HOMO level lies at higher binding energies makes it more convenient for analysis, since the HOMO level is much easier to separate from the emission of the V_O states.

8.2.2 Titanium

The Ti 2p core level is a good indicator for the dye-substrate interaction as well. The spectra have been taken at $h\nu = 600\,eV$ and normalized to the maximum of the main emission in order to exclude the damping by adsorbed dye.

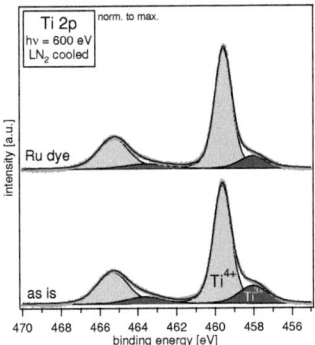

Figure 8.8: The Ti 2p core level emission measured at $h\nu = 600\,eV$.

Like the oxygen vacancies (Figure 8.7 right), the intensity of the Ti^{3+} states decreases upon dye adsorption (Figure 8.8). The Voigt fit reveals a quenching of about 32 % of the integrated intensity of the Ti^{3+} states. This evidences that the dye anchors at non saturated titanium sites of the substrate like the solvent molecules do (Section 7.1.2). Thus the dye interacts with oxygen defects of the substrate.

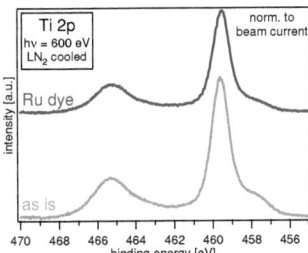

Figure 8.9: The Ti 2p signal of the uncovered substrate and the attenuated signal by the coverage of the adsorbed N3 dye.

By means of the Ti 2p substrate signal attenuation (Fig. 8.9), which is around 67 %

of the integrated intensity, the coverage of the substrate by the adsorbed dye has been estimated using Equation 3.9 analogous to the estimations of solvent coverages in Section 7. Assuming an inelastic mean free path of electrons λ_e of 6.9 Å according to Tanuma et al. [278], the coverage amounts to 2.8 Å. This value is much smaller than the dimensions of the molecule, which is around 10 Å in size according to Shklover et al. [149]. According to this estimation only a fraction of a monolayer of 28 % of dye is adsorbed on the surface. But it has to be noted, that the assumption of having a dense dye layer adsorbed on the TiO_2 substrate is most likely not correct, because the dye molecule has a very bulky shape. Therefore the estimated dye coverage value may be misleading, and the actual coverage may be higher than this estimated value. Nevertheless this estimation clearly suggests, that we do not exceed monolayer coverage.

8.2.3 Work function

The secondary edge for the untreated TiO_2 substrate and with adsorbed dye has been taken at $h\nu = 50$ eV, and a bias of 10 V has been applied (Figure 8.10). The onset of the secondary edge shifts down upon dye adsorption by 0.6 eV, i.e. the work function increases.

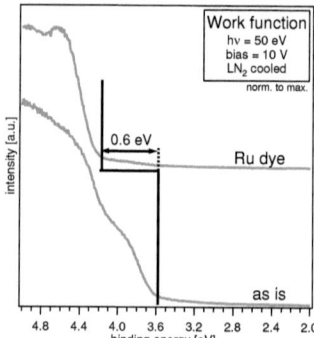

Figure 8.10: The work function for the untreated and dye sensitized TiO_2 substrate.

Since the dye adsorbs onto the surface upon deprotonation of the anchoring carboxyl groups [165], a surface dipole is created between the protonated and positively charged substrate surface and the negatively charged carboxyl groups [21, 81]. Due to that dipole, which points away from the surface, photoemitted electrons are retarded and consequently the work function increases.

8.3 Discussion of the results

By means of Synchrotron Induced Photoemission Spectroscopywe could characterize the dye monolayer with highest surface sensitivity with respect to its chemical components. The quantification by applying Voigt fitting procedures onto the measured spectra shows that the relative concentrations of the chemical components match reasonably to the respective number of atoms in the molecular structure, whereas the estimated concentrations for the C–C and O–C components suffer from ethanol solvent contamination.

The adsorption experiments of the dye onto the substrate show clearly that the dye molecules interact with the V_O states of the TiO_2 substrate. The dye adsorption is assumed being crucial for the device operation in two ways: At first it is necessary that the dye anchors onto the substrate to ensure an efficient charge transfer from the LUMO of the dye into the conduction band of the TiO_2 substrate. Forming a chemical bond leads to a large electronic overlap according to Fermi's Golden Rule (Section 1.3.1). Secondly the chemical surface bond formation reduces the amount of active oxygen vacancies, which act as recombination centers.

A representation of the energy alignment of the HOMO relative to the substrate based upon the gap region spectra (Fig.8.2.1) is discussed below in Section 9.4, page 138. Also a comparison to literature data is provided there.

9 TiO$_2$/N3 Ru dye/solvent interface

In this chapter the results of solvent coadsorption onto the dye sensitized TiO$_2$ surface are presented. In the first part the focus is on the interaction of the coadsorbed solvent with the substrate, whereas the second part emphasizes on the dye–solvent interaction. The chapter closes with the conclusions, where models are discussed in order to explain the observations during coadsorption.

9.1 Interaction of the coadsorbed solvent with the substrate

Ti 2p In order to investigate the interaction of the coadsorbed solvent with the dye sensitized substrate, spectra of Ti 2p at an excitation energy of $h\nu = 600$ eV have been recorded (Fig. 9.1 top left and right). The spectra have been normalized to the maximum at 459.6 eV in order to exclude damping by solvent coverage.

Comparing the intensity decrease of the Ti^{3+} states of the Ti 2p photoemission spectra of acetonitrile coadsorption onto the dye sensitized anatase substrate (Fig. 9.1 bottom), we observe different changes compared to the acetonitrile adsorption onto the untreated substrate: The coadsorbed acetonitrile quenches the Ti^{3+} states to a much smaller amount than the adsorbed acetonitrile onto the bare substrate. On the other hand the benzene coadsorption shows a similar trend like in the adsorption experiment. In other words, the dye seems to have a substantial influence on the intensity development in the case of the acetonitrile coadsorption, whereas for the benzene coadsorption this does not apply.

After the initial quenching of the Ti^{3+} states by the Ru dye, which is displayed by Figure 8.8 (page 123), both types of coadsorbed solvents cause a further quenching of the Ti^{3+} states. Since we assume submonolayer coverage of the dye on the surface (page 118, Section 8.1.3) the further quenching of the Ti^{3+} states is not surprising. In addition, the dye monolayer is not a dense layer due to the bulky geometry of the dye molecule. Thus the much smaller solvent molecules are expected to penetrate the dye layer easily.

Comparing the amount of quenched Ti^{3+} states depending on coverage, acetonitrile quenches the Ti^{3+} states to a similar extend (down to 65 % of initial intensity at 2.9 Å) as benzene (down to 73 % at 3.1 Å), as shown by Figure 9.1 right. This is in contrast to the conclusion drawn in Chapter 7 that acetonitrile interacts much stronger with the TiO$_2$ substrate than benzene does. Apparently the dye molecules prevent an interaction of acetonitrile molecules with the TiO$_2$ substrate (see Section 9.2).

128 CHAPTER 9. TIO₂/N3 RU DYE/SOLVENT INTERFACE

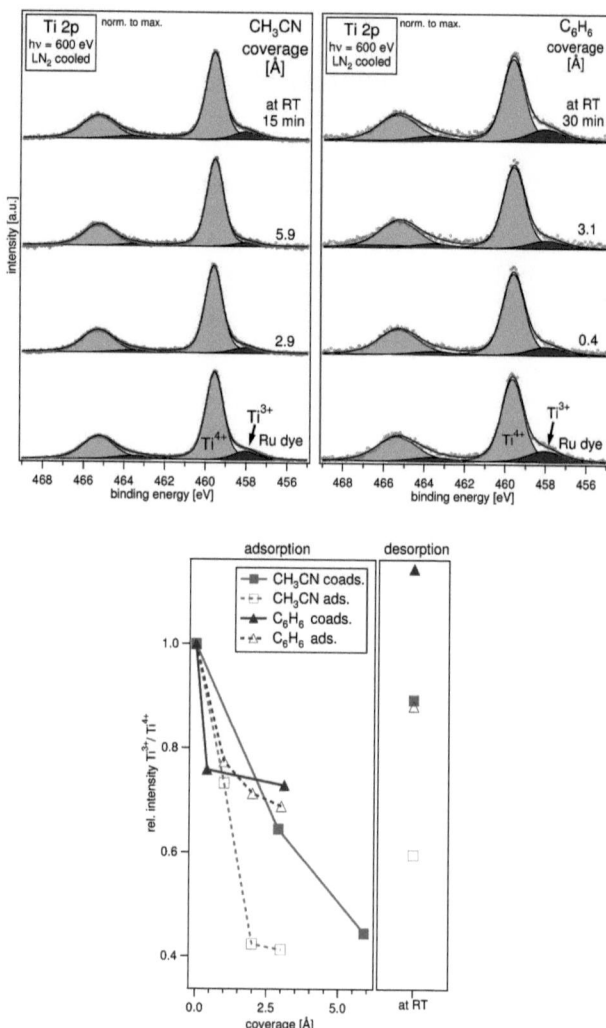

Figure 9.1: The Ti 2p emission line taken at $h\nu = 600\,\text{eV}$. **top left:** Spectra changes of the coadsorption of acetonitrile onto the dye sensitized TiO_2 substrate. **top right:** Coadsorption of benzene. **bottom:** Intensities of the Ti^{3+} states compared to the Ti^{4+} states as a function of estimated solvent coverage. For comparison the intensity changes of solvent adsorption onto the bare TiO_2 substrate from Figure 7.3 is also shown (empty dots, dotted lines).

9.2 Dye–solvent interaction

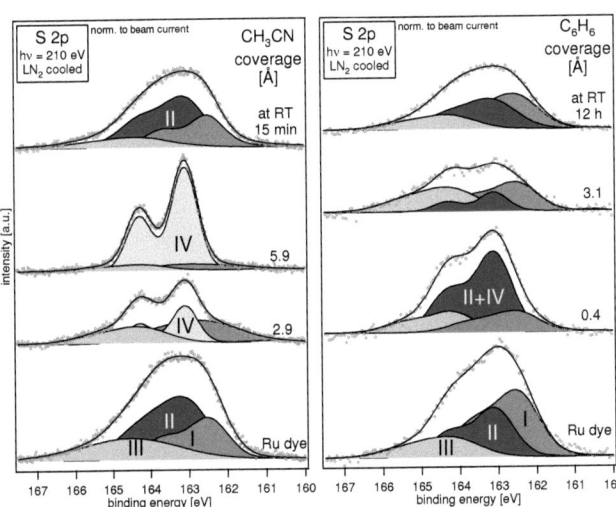

Figure 9.2: The S 2p core level. **left:** The spectra course of the coadsorption of acetonitrile. **right:** Coadsorption of benzene.

S 2p core level spectra The interaction of acetonitrile or benzene with the N3 dye adsorbed on the anatase surface can be followed by the relevant core level emissions of the dye. Sulphur is the most characteristic atom being exclusively present in the dye molecule. S 2p emissions for both coadsorptions have been measured at $h\nu = 210$ eV providing highest surface sensitivity (Fig.9.2). The S 2p spectra have been normalized to the beam current.

Each spectral development starts with a dry dye layer, which shows a broad emission. Looking at the coadsorption of acetonitrile remarkable changes in the spectral features are displayed (Fig.9.2 left). With the first coadsorption step of acetonitrile the emission line changes drastically regarding its shape and intensity. The next adsorption step reveals very interesting and astonishing peculiarities: Although the coverage of adsorbed acetonitrile is higher than in the first adsorption step, the intensity of the S 2p line is increased enormously accompanied with a further decrease of the line width. Upon desorption of CH_3CN we get a broad S 2p emission line similar to that before coadsorption of the solvent. Therefore we conclude the coadsorption of the solvent is a reversible process.

Inspecting the S 2p line of benzene coadsorption (Fig.9.2 right) such remarkable changes of the S 2p emission line are not observed. Upon adsorption the S 2p line is just damped in intensity. The reversibility of the benzene coadsorption is obvious from the spectral changes with annealing to room temperature.

For having a closer look at these interesting features we have performed a fit including three doublet Gauss-Lorentz (Voigt) profiles (spin-orbit splitting of 1.2 eV, intensity ratio of 1:2) in order to deconvolute the different chemical components (Fig.9.2 left), which are located at binding energies of 162.5 eV (I), 163.3 eV (II) and 164.2 eV (III) respectively, as have been depicted in Section 8.1.4 already: Component I represents the thiocyanate groups, which are isolated from each other, and point away from the substrate. The second component II represents the fraction of interacting thiocyanate groups, either with other dye molecules or with the substrate. The third component III remains unclear.

In the case of acetonitrile coadsorption a component IV appears ($E_B = 163.1$ eV), which is located at slightly lower binding energy as component II, whereas the remaining components I and III almost disappear. Apparently, component II disappears completely upon coadsorption of acetonitrile. Since the decrease of component I coincidences with the rise of component IV, we assign this component to solvated molecules, which do not interact with the substrate. In other words, component IV originates from component I, which is shifted 600 meV to higher binding energies due to solvation. Because the sulphur atoms holds two lone paired electrons, the positive end of the solvent molecule is oriented towards the thiocyanate group. Thus the potential in the near vicinity of the sulphur atom is decreased, which leads to a shift towards higher binding energy.

The assignment of the components I and II made above are justified by the following argument: If the sulphur atom interacts with the substrate, most likely it interacts with the reduced Ti^{3+} states. In that way the sulphur atom acts as an electron donor and the Ti^{3+} as an electron acceptor forming a higher oxidation state (Ti^{4+}) after transferring its electron into the conduction band.

Since the coadsorbed solvent reduces the amount of Ti^{3+} states as we found above, the dye molecules have less sites left to interact with the substrate via the thiocyanate groups. Hence the component IV is strongly increasing in intensity with consecutive adsorption, while component II disappeared. The same trend has been observed by Karlsson et al. [294] using CuI as the coadsorbant. Karlsson has deconvoluted the S 2p into three components for the CuI coadsorption, but with higher binding energy shifts between component I and II, i.e. he found 0.9 eV instead of our value of 0.6 eV.

Regarding the acetonitrile coadsorption, the observations described above can be explained as follows: Concerning the first adsorption step the slight sharpening of the spectra is due to a (partial) solvation of the dye molecules by the surrounding solvent. The decreasing intensity indicates that the solvent molecules lies on top of the dye layer. In the second adsorption step acetonitrile penetrates the dye molecule film. As a result of the solvation the thiocyanate groups are isolated from each other, which is observed by the drastic sharpening of the S 2p line. This solvation process is visualized in a cartoon in the conclusions of this chapter (Section 9.4, Figure 9.10).

Unlike acetonitrile, benzene as an unpolar solvent does not solvate the negatively charged thiocyanate group in a pronounced way. Although in the first adsorption step the intensity of component II increases slightly, which we attribute to a partial solvation of the dye molecules (component IV). In the second adsorption step benzene mainly damps the S 2p emission. Hence the solvent molecules just stay on top of the monolayer.

9.2 Dye–solvent interaction

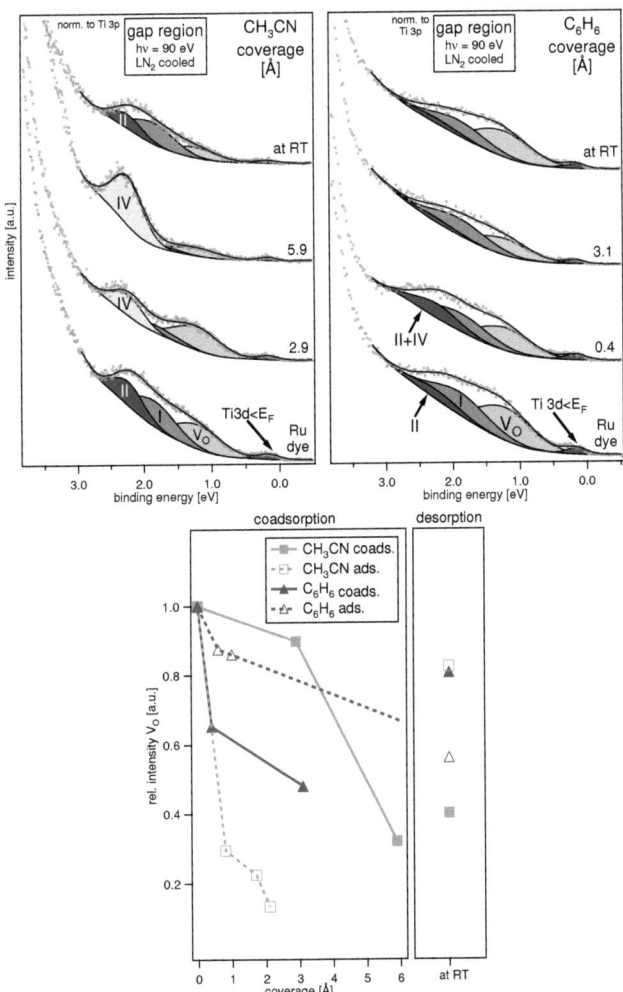

Figure 9.3: The gap region recorded at $h\nu = 90$ eV, showing the HOMO of the dye and the Ti 3d states of the substrate. **top left:** Acetonitrile coadsorption. **top right:** Benzene coadsorption. **bottom:** The intensity development of the V_O states for both acetonitrile and benzene coadsorption. For comparison the intensity of solvent adsorption onto the bare TiO$_2$ substrate from Figure 7.4 is also shown (empty dots, dotted lines).

Ti 3d / HOMO valence band spectra The gap region spectral changes are of special interest (Fig. 9.3), because here one can actually see the interaction of both substrate and dye features with the coadsorbed solvent all at once. The spectra have been taken at $h\nu = 90$ eV and normalized to the Ti 3p emission for eliminating the damping.

One can observe a decrease of V_O state intensity upon coadsorption of acetonitrile and benzene, respectively, parallel to the Ti 2p spectral effects. Whereas acetonitrile quenches the V_O states almost completely in the second adsorption step, the benzene coadsorption does not change the V_O intensity that much. Nevertheless, as for the Ti^{3+} states (Figure 9.1 top left) the interaction of coadsorbed acetonitrile with the V_O states is completely different to the observations after adsorption onto the untreated substrate (Fig.9.3 bottom, dotted line). In the case of coadsorbed acetonitrile the V_O states are much less quenched. For the benzene coadsorption the decrease of intensity follows a similar trend as for the adsorption (dotted line).

A deconvolution of the gap emission lines throughout the spectra into four Voigt profiles has been performed to analyze the spectra into their respective contributions. Since the HOMO level on the N3 molecule is mainly located at the thiocyanate group [9,151,152,293] involving the sulfur atom, the fit procedure is similar to the performed fit of the S 2p emission line. We divided the HOMO level into two Voigt profiles I and II, which fit surprisingly well into the measured spectra. The first profile I lying at BE = 1.8 eV displays the non-interacting fraction of thiocyanate groups. The second profile II located at 400 meV higher binding energy (BE = 2.2 eV) is attributed to the molecule fraction, that interacts with the anatase substrate. Qualitatively we find the same trend for the acetonitrile and benzene coadsorption as observed in the S 2p spectra, where the shift between component I and II was 0.6 eV.

Upon coadsorption of acetonitrile the HOMO level I is drastically decreased in intensity. According to the S 2p spectra (Fig. 9.2 left), component IV is assigned to the solvated dye molecules. Component IV appears after the first acetonitrile adsorption step and even increases in intensity after the second adsorption step, whereas component II disappears. These attributions are justified by the fact, that component IV is more narrow than the component II (Table 9.1), which is according to the observations we made at the S 2p core level emission spectra (Fig. 9.2 left).

Table 9.1: The full width half maximum for components II and IV during acetonitrile coadsorption (Figure 9.3) top left.

experimental step	component	FWHM
Ru dye	II	0.37
2.9 Å CH_3CN	IV	0.33
5.9 Å CH_3CN	IV	0.29
at RT	II	0.40

In the case of benzene coadsorption the spectra lead only to small changes in intensity and shape of the HOMO emission (Fig.9.3 top right). Interestingly the intensity of the V_O states after the first acetonitrile coadsorption step stays almost the same. After the subsequent second coadsorption step the intensity is drastically reduced. In the case of benzene the intensity decreases only moderately.

A possible explanation for the different quenching behaviour of coadsorbed acetonitrile is given in Figure 9.10. At first, acetonitrile interacts with the dye. Then with increasing dose the solvent molecules penetrate the dye layer to the substrate underneath, where they quench the V_O states. In contrast, the nonpolar benzene molecules do not solvate the polar

9.2 Dye–solvent interaction

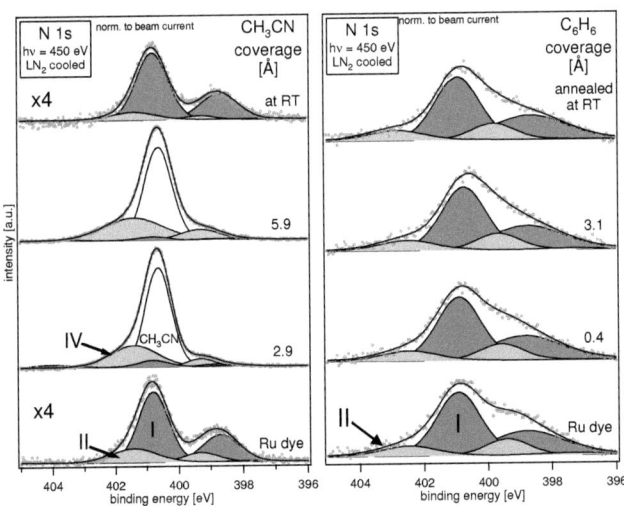

Figure 9.4: The N 1s emission line taken at $h\nu = 450$ eV. **left:** The spectra course for the acetonitrile coadsorption. For better visibility, the intensity of the dry dye emission and the emission taken after annealing have been multiplied by the factor of 4. **right:** The benzene coadsorption.

dye molecules as efficiently. The benzene molecules either just cover the dye layer without strong interaction or penetrate the dye layer and interact with the substrate on bare areas.

It has to be noted, that the starting situations for the coadsorption experiments differ slightly. As already observed in the S 2p line, the gap region spectra of the dry dye of the acetonitrile and benzene adsorption series are different in shape. But the fact, that the difference in coadsorption behavior of either acetonitrile or benzene is so pronounced, allows us to draw clear qualitative conclusions.

N 1s core level spectra The amount of acetonitrile coadsorbed onto the dye sensitized substrate can be inferred from the N 1s emission line (Fig.9.4 left). The N 1s emission has been measured at $h\nu = 450$ eV for highest surface sensitivity and normalized to the beam current. The acetonitrile emission is located at 400.6 eV not changing its binding energy for both coadsorption steps.

The coadsorption of acetonitrile onto the dye sensitized anatase substrate can directly be deduced from inspecting the N 1s spectrum. The signal of the dry dye is split into two main peaks. The smaller peak at lower binding energy is attributed to the nitrogen in the thiocyanate group and the larger peak at higher binding energies is assigned to the bipyridine group.

As seen by an applied Voigt fit we observe a binding energy shift of the dye emissions upon coadsorption of acetonitrile towards higher values, but no damping due to acetonitrile coverage. To support this results observed in the N 1s spectra, we also applied a Voigt fit deconvoluting the Ru dye emission line into two doublets (I and II) of thiocyanate and bipyridine nitrogen atoms, respectively. The binding energy difference between both doublets is 0.6 eV, and the energy distance between the thiocyanate and the bipyridine contribution is 2.1 eV as already mentioned in Section 8.1.2. We are aware of the splitting especially within component pair II concerning the intensity of the bipyridine contribution, which might be caused by a contribution of acetonitrile emissions, which cannot be separated clearly. In agreement to both S 2p and the HOMO signal described above we assign the signal pair I to the fraction of non solvated dye molecules, whereas signal pair IV is attributed to the solvated dye.

Following the spectra throughout the acetonitrile coadsorption one can see an increase of the ratio of component pair II to pair I. Furthermore the overall intensity of the dye signal emission increases despite of the ongoing coadsorption of acetonitrile, which supports the above given interpretation of both S 2p and HOMO level emissions. In contrast the adsorption of benzene does not show any remarkable change in the spectral development (Fig.9.4 right).

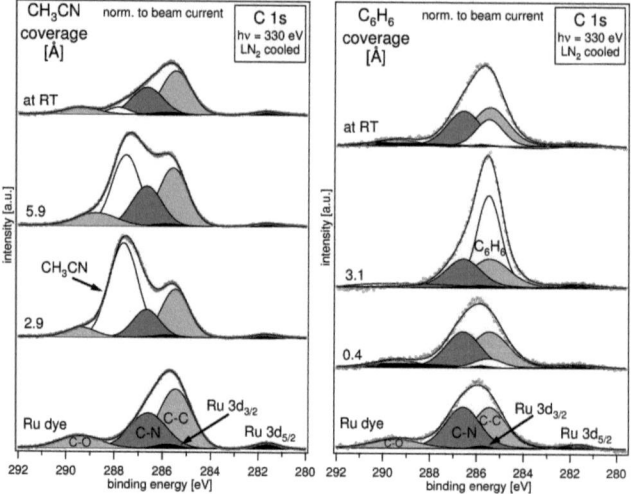

Figure 9.5: The C 1s and Ru 3d orbitals, measured at $h\nu = 330$ eV. **left:** C 1s of the acetonitrile coadsorption. **right:** C 1s of the benzene coadsorption.

C 1s core level spectra Although the C 1s signal (Fig.9.5 left top and bottom) suffers from carbon contamination of the substrate (Section 8.1.1), it supports the observations stated above. The signal was taken at a photon excitation energy of $h\nu = 330$ eV for highest possible surface sensitivity and normalized to the beam current.

Again we performed a fit procedure including three Voigt profiles to the C 1s emission line

9.2 Dye–solvent interaction

and one doublet accounting for the contribution of the Ru 3d orbital (spin orbit split = 4.2 eV) [239]. Actually the fit should contain separate profiles for each component as realized for the other emissions presented above representing the different interactions of the respective component (interacting, non-interacting with substrate, and a third component, which is not clearly assigned). But for the C1s emission, which includes three carbon components and a Ru 3d doublet, it is not feasible to produce a reliable fit by splitting each of these chemical components into three subcomponents I, II and III as performed for the other core levels S 2p and N 1s. The carbon components are assigned accordingly to Liu et al. [292] to aromatic carbon (C–C) at BE 285.5 eV, C–N carbon at BE = 286.7 eV and carboxyl carbon at 289.4 eV. Additionally, the Ru $3d_{5/2}$ is located at 281.6 eV (refer to Section 8.1.1). We have to note that in particular the profile assigned to aromatic carbon (C–C) contains a not negligible amount of contamination present on the substrate as it has been already shown by Figure 9.5. Nevertheless the remaining profiles except the C–C contribution match reasonably with the number of the atoms in the molecule (see Table 8.2). It also has to be noted that the acetonitrile contribution has been fitted by one component only, which is of course an oversimplification, since an acetonitrile molecule consists of two carbon atoms with different chemical environment. The intensity changes of the C–C and C–N contributions of the dye molecule are listed in Table 9.2.

	C–C	C–N
E_B [eV]	285.5	286.7
I_{drydye}	1	1
$I_{1^{st}coads.}$	0.67	0.57
$I_{2^{nd}coads.}$	0.87	0.91
$I_{desorb.}$	0.66	0.65

Table 9.2: The integrated intensities of the C–C and the C–N component during coadsorption of the solvent, normalized to the intensities for the dry dye layer.

The numbers in the table allows us to underline the observations discussed above: In the case of acetonitrile coadsorption the first coadsorption step initially shows a damping of the C–C and C–N profiles belonging to the dye molecules upon acetonitrile coadsorption (acetonitrile emission at BE = 287.6 eV). With the second step of increasing acetonitrile adsorption the acetonitrile signal surprisingly decreases again, whereas the C–C and C–N carbon components are increased in intensity, which supports our model of the penetration of acetonitrile into the dye monolayer.

The C 1s signal is the appropriate photoemission line to monitor the coadsorption of benzene (Fig. 9.5 right). We observe a strong signal increase at BE = 285.5 eV due to coverage with benzene solvent molecules. In analogy to the acetonitrile spectra, the fit shows three carbon emission lines and one Ru 3d doublet. The dry dye emission is in a slightly different starting situation than for acetonitrile as the intensity ratio between C–C and C–N carbon atoms disagree with the one in the acetonitrile spectral series. This may be due to a missing differentiation between the components I, II and III. In addition the quantitative analysis suffers largely from the fact that in the benzene contribution both C–N and C–C components are superimposed. Unlike in the case of acetonitrile coadsorption, a quantitative statement about the intensity changes during coadsorption is not possible.

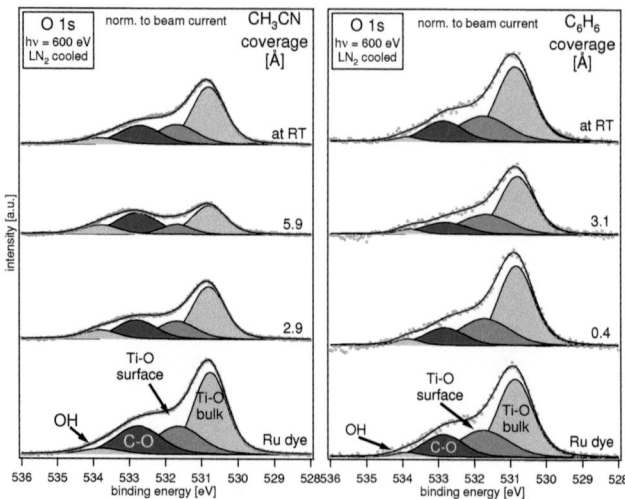

Figure 9.6: The O 1s core level emission line ($h\nu = 600\,\text{eV}$) depending on solvent coadsorption. **left:** The coadsorption of acetonitrile. **right:** The coadsorption of benzene.

O 1s core level spectra The O 1s emission line has been measured with an excitation energy of $h\nu = 600\,\text{eV}$ and normalized to the beam current.

Qualitatively inspecting the acetonitrile series (Fig. 9.6 left) shows, that the shoulder at higher binding energy stays at the same intensity, whereas the substrate emission is damped due to solvent coverage. In contrast, benzene coadsorption shows an equal decrease of intensity for both shoulder and substrate oxygen peak.

The deconvoluting fit of the photoemission spectra has been performed by four components being assigned to bulk oxygen, surface oxygen, carboxyl groups (protonated and unprotonated) and adsorbed hydroxyl groups (according to Section 8.1.3, Fig. 8.5). The carboxyl component indicates the adsorption of the dye onto the TiO_2 substrate, while all other components belong to the substrate. The intensity of the substrate emissions (Ti–O bulk and Ti–O surface component) is damped by the coadsorbed acetonitrile, whereas the C–O component is almost not decreased in intensity. Karlsson et al. [294] observed the same trend regarding the dye / substrate emission intensity ratio upon coadsorbing CuI onto the dye sensitized substrate as compared to our acetonitrile coadsorption. In contrast to acetonitrile benzene damps both substrate and dye emissions to the same degree.

Also the O 1s photoemission line supports our model of the interpenetration of the dye monolayer by acetonitrile molecules as well as the remaining on top of the dye monolayer for the benzene molecules (see discussion of this chapter, Section 9.4).

9.3 Work function changes

We have already discussed the work function changes for the dye adsorption and the adsorption of solvent separately, but now we would like to investigate the trend of work function for the coadsorption of dye and solvent. The secondary edge measurement has been taken at an excitation energy of $h\nu = 50$ eV, and a bias voltage of -10 V has been applied (Fig. 9.7).

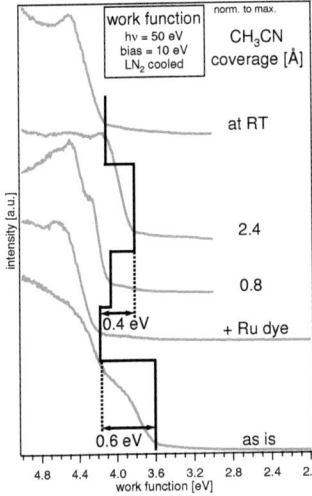

Figure 9.7: The change of work function during the adsorption of dye and the coadsorption of acetonitrile.

Upon dye adsorption the work function increases by 0.6 eV as it has been shown by Fig. 8.10 already. By coadsorption of acetonitrile this trend is reversed. By increasing the dose of acetonitrile we observe a shift of 0.4 eV towards lower work functions. This trend has been observed for the adsorption of acetonitrile onto the untreated substrate as well, but to a slightly different extent. Upon desorption the work function is the same as for the dye adsorbed surface.

A possible explanation of this trend is discussed in the context of dye reorientation in the discussion section below (Section 9.4).

9.4 Discussion of the results

Electronic alignment The alignment of the energy levels of the dye with respect to the TiO_2 substrate are of prime importance for device function. Based upon the experimental results of acetonitrile coadsorption from the gap region spectra an energy scheme of the substrate/dye interface has been derived, shown in Figure 9.8.

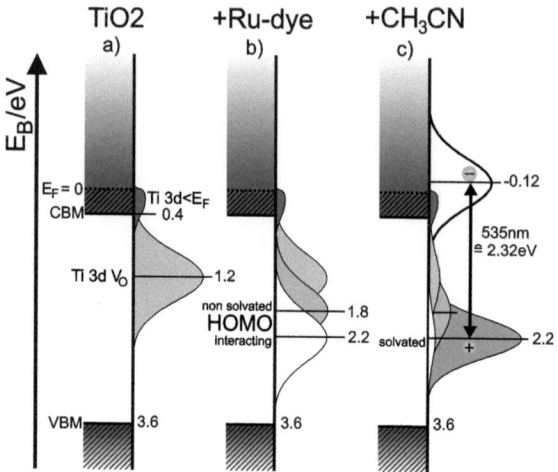

Figure 9.8: Energy scheme of the TiO_2/dye/solvent interface displaying all the relevant energy levels of all species involved. The LUMO of the dye has been estimated using the maximum of optical absorption of the dye at 535 nm (2.32 eV [20]).

As already described in detail in Section 8.2.1 (page 121), the V_O states ($E_B = 1.2$ eV, Figure 9.8 a)) are quenched partially upon dye adsorption (Figure b)), since the dye is bound to the substrate via carboxyclic oxygen. In contrast the $Ti3d < E_F$ states almost keep their initial intensity. For the dye HOMO level we consider two different components, according to Section 8.2.1, i.e. an interacting one, which is located at $E_B = 1.8$ eV, and an isolated HOMO, which is about 400 meV at higher binding energy.

Upon coadsorption of acetonitrile, we observe an additional quenching of the V_O states. In contrast the $Ti3d < E_F$ states are not decreasing in intensity to that large extent. Following arguments of Gregg et al. [17, 80] we assume, that the $Ti3d < E_F$ states are beneficial for the device component by forming a conduction channel for electrons. The dye and the solvent do not change these gap states significantly.

Concerning the HOMO of the dye the interacting component disappears upon coadsorption of acetonitrile. This result has been confirmed by the S 2p emission line, where we also observed that interacting component disappears (Fig.9.2). We assume, that this has a number of very significant consequences with respect to the energetic alignment and for the dye orientation.

9.4 Discussion of the results

In order to estimate the energy alignment of the LUMO level of the dye, the maximum of optical absorption of the dye of 535 nm [18] ($h\nu = 2.32$ eV) has been plotted into the energy scheme. The onset of the LUMO is drawn in accordance with the onset of optical absorption at 780 nm [18] ($h\nu = 1.6$ eV). One has to consider, that a direct comparison of energy values between optical and photoemission spectroscopy is not possible, since in optical absorption a *Frenkel exciton* on the dye molecule is created. This means that the photohole and the excited electron remain on the same dye molecule, which are bound together by electrostatic interactions, the exciton binding energy. Usually an exciton binding energy of about 0.6 eV [19] has to be considered. It is assumed that the exciton binding energy affects the hole in the HOMO and the electron in the LUMO level by the same amount. But as the acetonitrile has a high dielectric constant ($\epsilon = 36.6$ at 20°) it reduces significantly the exciton binding energy (proportional to $1/\epsilon^2$ according to the Bohr model). Therefore we do not consider any strong shift of the HOMO and LUMO level. With these assumptions the HOMO level is located at $E_B = 2.2$ eV and the LUMO at 0.12 eV above the Fermi level. The onset of the LUMO is located at 0.5 eV binding energy.

The non interacting and non solvated HOMO is located at $E_B = 1.8$ eV. If these levels were relevant for charge separation, the LUMO level would be located even 0.52 eV above the Fermi level. As a consequence the photoinjection of electrons would be favored, but on the other hand, the rereduction of the dye would be even slower since the HOMO level is moving towards the redox potential of the electrolyte.

In order to compare our binding energy value of the solvated HOMO level ($E_B = 2.2$ eV) measured by Photoelectron Spectroscopy with literature data determined by electrochemical measurements [18] (Fig.1.1, page 15), one has to take the work function of the nanocrystalline anatase substrate into account, which is in our case $\phi \sim 3.7$ eV. This value agrees to the value obtained by the Ultraviolet Photoemission Spectroscopy data of Orendorz et al. for pristine nanocrystalline anatase ($\phi = 3.7$ eV) [140]. Since the electrochemical measurements by Nazeeruddin et al. [18] have been conducted in acetonitrile solution, we compare the solvated component of the HOMO level only. An ionization potential of 5.5 eV for the HOMO level as obtained from current-voltage curves leads after substraction of the work function $\phi = 3.7$ eV to a binding energy of the HOMO level of $E_B = 1.8$ eV.

This value deviates by 400 meV from our value of $E_B = 2.2$ eV, but matches to the value of the non solvated HOMO component. As a reason for the deviation the uncertainty of the work function value must be considered, which was different from sample to sample. Another reason for the deviation may be the different preparation procedure performed by Nazeeruddin et al. in comparison to our dye adsorption experiments. While we adsorbed the dye for 5 min only, Nazeeruddin et al dipped it in dye solution for several hours. Since the amount of adsorbed dye seems to vary by the adsorption time [159], we assume that Nazeeruddin et al. exceeded monolayer coverage, which probably leads to a bigger fraction of non solvated dye molecules.

Orientation of the dye Based on the intensity variations, e.g. the increase of both S 2p emission lines (Fig.9.2 left) and the HOMO level (Fig.9.3 top left) upon coadsorption of acetonitrile, we conclude, that the acetonitrile coadsorption leads to a reorientation of the dye molecules (Figure 9.9). As we have seen by the Ti 2p spectra (Fig.9.1 left) and the gap region (Fig.9.2 left), we know by the quenching of both Ti^{3+} and V_O states in the second coadsorption step, that acetonitrile interacts with the TiO_2 substrate. Since the nitrogen atom of acetonitrile is expected to interact with the Ti^{3+} sites, the interaction of acetonitrile competes with the interaction of the sulphur atoms belonging to the dye molecule. As a consequence, the molecule is changing its orientation to two possible adsorption geometries, which are illustrated by Figure 9.9. The third possible adsorption geometry (compare Figure 2.12 d)) does not allow an interaction of sulphur on a flat surface, therefore it is not discussed here.

In the case of benzene coadsorption, no orientation change is evident from the S 2p and the HOMO level. Neither the S 2p nor the HOMO level show intensity enhancements, but the intensity is reduced continuously for both emission lines. Nevertheless we observe an interaction with the substrate, since both Ti^{3+} and V_O states are quenched slightly. However, it is clearly evident, that for benzene coadsorption no orientation changes take place.

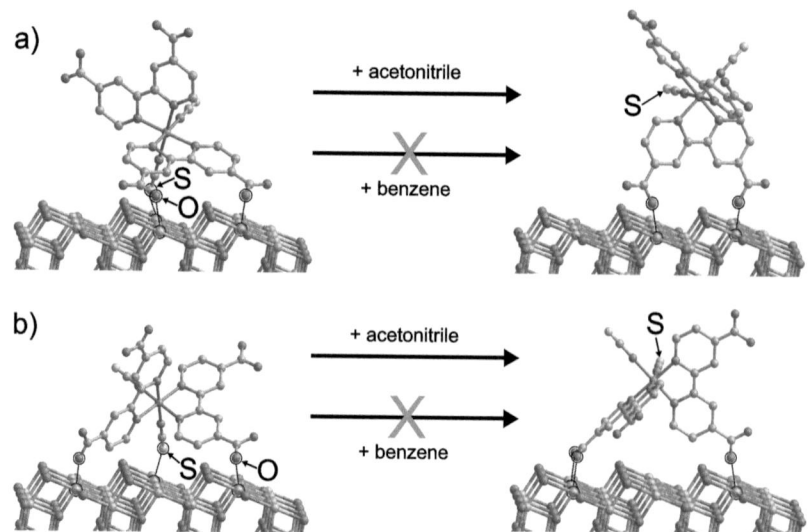

Figure 9.9: Two possible reorientations of the N3 dye molecules upon solvent coadsorption. The N3 dye molecule and the anatase (101) surface are drawn to scale according to [149, 160]. The arrows indicate, that the reorientation only takes place for acetonitrile coadsorption, but not in the case of benzene coadsorption. The sulphur (marked with S), that initially interacts with the substrate, points away from the surface after reorientation. **a)** This reorientation does not involve the formation of another chemical bond to the substrate via carboxyclic oxygen (marked with O). **b)** This reorientation leads to formation of another bonding between carboxyclic oxygen and substrate.

On a more macroscopic scale than individual dye molecules, which concerns the complete

9.4 Discussion of the results

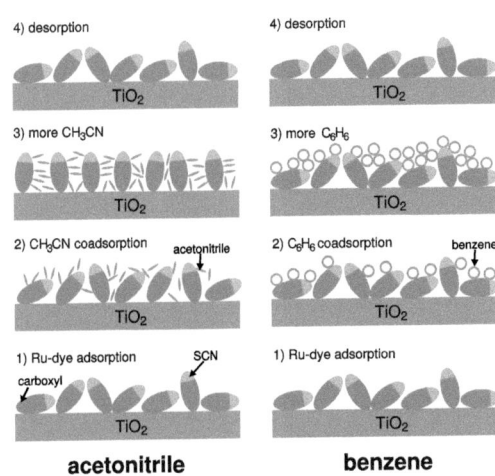

Figure 9.10: A schematical representation of the processes during the coadsorption of the solvents acetonitrile (left) and benzene (right) onto dye covered TiO$_2$.

dye monolayer, a simple geometric model (Fig.9.10 left) can be deduced from the photoelectron spectra and the trend of the work function [295]: In the absence of solvent the N3 sensitizer molecules lie disordered on the TiO$_2$ surface with their NCS groups interacting with the surface and neighboring sensitizer molecules causing statistical broadening of the S 2p level. Coadsorbed acetonitrile molecules evidently penetrate the sensitizer layer, separate the molecules from each other and force the N3 molecules to point with their NCS groups away from the substrate, thereby reducing the statistical broadening. Thus acetonitrile serves not only as a medium, in which the hole transporting redox species can move. But it also has a distinct and important influence on the adsorption geometry. In comparison to the benzene coadsorption (Fig.9.10 right), which does not show similar effects, we strongly suggest, that the effect as observed with acetonitrile depend on the polarity of the solvent molecules.

This model of the reorientation of the dye upon solvent coadsorption is confirmed by the trend of the work function shown by Fig.9.3, page 137. In order to explain the change of the work function to lower values (by 0.4 eV from the dry dye layer to 2.4 Å of acetonitrile coadsorption) we need to separate the effects of dye and solvent in terms of the formation of a surface dipole (Fig.9.11).

The dye molecule consists of two opposite dipoles, both pointing away from the positive central atom. The two bonding carboxyl groups (we assume, two carboxyl groups are bound to the surface [161]) have three lone paired electrons and one bonding pair each (thus about eight electrons), but the two thiocyanate groups possess only four lone pair electrons (Fig.9.11 a)). In addition, according to Srikanth et al. [151], the distance between the ruthenium center atom and the carboxyl group amounts to about 6.5 Å, whereas the thiocyanate groups are only about 4.9 Å away from the metal center atom. Thus the dipole from the center of the dye towards the carboxyl groups is larger than in direction to the thiocyanate group. Since the dipole moment is increasing linearly with the distance of the opposite charges (like a plate capacitor), the overall dipole is pointing towards the carboxyl group (Fig.9.11 b)).

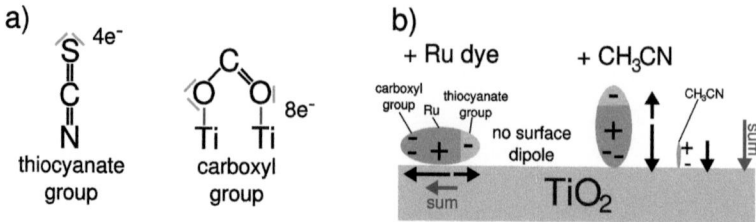

Figure 9.11: a) The number of lone pairs differ for the carboxyl groups and the thiocyanate groups respectively. b) In order to explain the work function trend (Fig.9.3), a dipole model for the coadsorption has been applied, showing the sum of the dipole for both situations.

Now we are able to develop a model, accounting for the surface dipole change, which results in the change of the work function (Fig.9.11 b)). For the dry dye layer, the molecules lie disordered on the surface as shown by Fig.9.10. By coadsorption of acetonitrile the work function shifts back by 0.4 eV. Now we have to take into account, that the dye layer covers only a fraction of a monolayer. The coverage amounts to 28 % according to our estimation by means of the attenuation of the Ti 2p main emission line (Section 8.2.2). We assume that the shift of 400 meV contains two contributions: The first one, to a fraction of 72 % (0.29 eV), is due to the surface dipole formation by the solvent molecules as described by Figure 7.6, page 105. The remaining contribution of the work function change of 0.11 eV we attribute to the change of the orientation of the dye with respect to the substrate surface. Consequently the dye has to be straightened up (Fig.9.11 b) right) according to the model described by Figure 9.9.

To evaluate these orientation effects with respect to the device functional, one has to consider the role of the dye. The dye is responsible for charge separation of holes and electrons via a vectorial charge transfer as described by O'Regan et al [1]. This means that the dye acts as an electronic membrane, which transfers electrons via the bipyridine groups (LUMO) into the conduction band of the TiO_2 substrate, whereas on the other sterical end of the dye a hole is injected into the thiocyanate groups (HOMO). Thus the orientation of the dye is of prime importance for device operation: The favorable geometry is that only the carboxyclic groups interact with the TiO_2 substrate. The thiocyanate groups should stick into the electrolyte like antennas, favoring the rereduction of the dye by the I^-/I_3^- redox couple.

Another important reason that this distinct orientation is of prime importance is related to recombination. Since the HOMO contains the photohole upon excitation by incident light, the direct contact with the TiO_2 substrate should be avoided. If the HOMO contacts the substrate, recombination path will be created, since photoinjected electrons in the TiO_2 conduction band will easily recombine with the holes in the nearby HOMO level. This recombination is probably favored by the remaining V_O defect level of the TiO_2 surface.

10 TiO$_2$ / iodide / solvent interface

In this chapter the adsorption of electrolyte species, forming a TiO$_2$/iodide interface, is elucidated. Two different iodide containing species have been used, besides LiI also 1-propyl-3-methylimidazolium iodide (PMII), which is an ionic liquid, denoted as molten salt in this chapter. In addition, the effect of coadsorbed acetonitrile is described for both adsorbed iodide compounds.

10.1 The two electrolyte components: LiI and molten salt

Valence band and I 4d spectra The adsorption of the electrolyte was performed from acetonitrile solution, according to Section 4.2.2, page 70. First of all it has to be mentioned that in this work only the reduced part of the I$^-$/I$_3^-$ redox couple, the I$^-$ species, has been coadsorbed. It is not possible to adsorb both species of the redox couple from solution (according to the preparation carried out in this work) due to technical reasons: In order to adsorb the I$_3^-$ species, besides the respective iodide salt, elemental iodine has to be added. The problem is, which could not be overcome, that iodine sublimes at normal conditions, and in vacuum even faster. Due to the volatility of iodine it is not easily possible to investigate it.

Moreover, the presence of the I$_3^-$ ions have been found by other groups to play a detrimental role in a way, that they form a I$_2$SCN$^-$ complex with the thiocyanate group of the dye [296, 297]. This complex formation can result in the loss of the thiocyanate ligand from the dye. Since this degeneration of the dye, induced by the I$_3^-$ ions, is not the scope of this work, the investigation of the complete redox couple has not been accomplished.

However, since the focus of this work is on the dye sensitized interface (working electrode), the reduced I$^-$ is of prime interest. The photooxidized dye can only be reduced by the I$^-$ redox species. In Chapter 11, the electronic alignment relative to the dye levels has been discussed based on literature.

Valence band spectra at $h\nu = 90$ eV were taken for both adsorbed LiI and PMII molten salt (Fig.10.1 left). One can clearly observe the I 4d emission line for both LiI and molten salt. In addition there is a Li 1s emission located at $E_B = 56.1$ eV.

3 drops of $\sim$0.01 M LiI and 10 drops of $\sim$0.005 M molten salt dissolved in acetonitrile have been adsorbed onto the TiO$_2$ substrate, respectively. Since the solvent (acetonitrile) evaporates during adsorption, the amount of adsorbed iodide should be similar on both samples. But as one can see from the valence band spectrum, the intensity of the I 4d of

CHAPTER 10. TIO$_2$ / IODIDE / SOLVENT INTERFACE

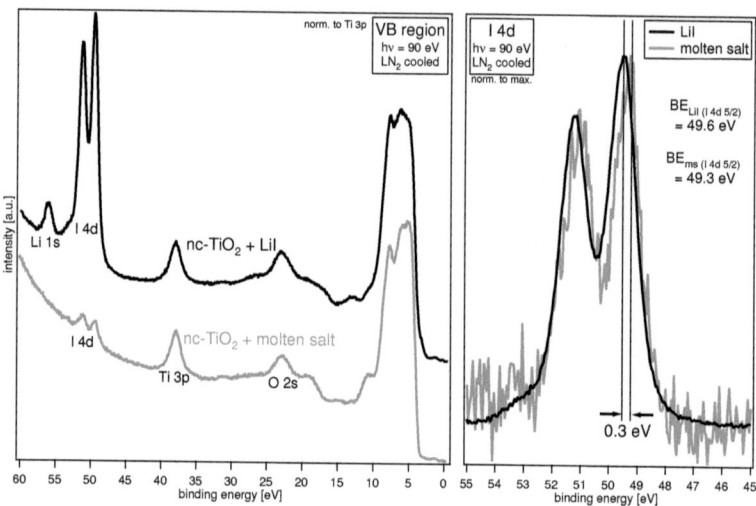

Figure 10.1: left: Valence band spectra of the molten salt (bottom) and the LiI adsorption (top). **right:** The I 4d emission for both molten salt and LiI compared with respect to their binding energy.

the LiI adsorption is much higher than of the molten salt. This drastic intensity difference between LiI and the molten salt adsorption can be (partially) explained: Because of the large PMII$^+$ cation of the molten salt ($r_{PMII^+} = 570$ pm) the spatial distribution of iodide is much less dense than in the case of LiI, where the lithium cation ($r_{Li^+} = 73$ pm) is much smaller than the iodide anion ($r_{I^-} = 206$ pm).

In the case of LiI the intensity of the I 4d emission line is much higher than the Li 1s emission. This depends on the different cross sections for both lines at $h\nu = 90$ eV (Li 1s: 2 mb, I 4d: 15 mb). In addition, the I 4d orbital shows a $I4d \rightarrow I5p$ resonance around an excitation energy of $h\nu = 90$ eV, which leads to the high intensity of the I 4d level [298–300].

While the binding energy of the I 4d emission of the LiI is $E_B = 49.6$ eV, the I 4d emission of the molten salt is located at 49.3 eV (Fig.10.1 right). This binding energy difference arises from the different cations of the salt. A model trying to explain the origin of the binding energy difference is given in the conclusion of this chapter below (Section 10.3). Also a comparison to literature data is given by Table 10.1.

I 5p core level spectra The electronic configuration of iodine ($Kr 4d^{10} 5s^2 5p^5$) indicates, that the highest occupied orbital is the I 5p orbital. The energetic alignment of the redox couple I^-/I_3^- is of prime importance for the device operation in terms of the redox potential of the electrolyte. Hence the investigation of the I 5p orbital is of special interest. In order to separate the I 5p level from the TiO$_2$ substrate valence band emission, difference spectra

10.1 The two electrolyte components: LiI and molten salt

has been calculated between the untreated substrate and adsorbed LiI (Figure 10.2 left). For better visibility the difference spectra have been multiplied by the factor of 5.

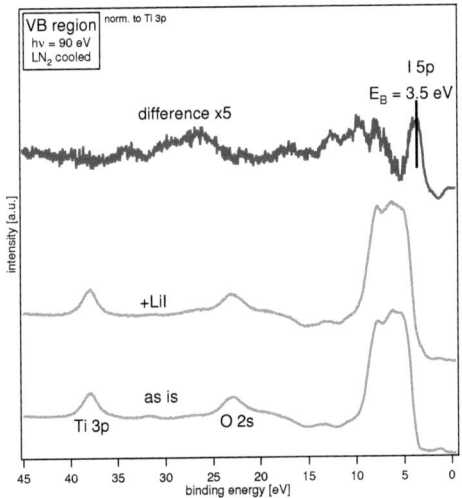

Figure 10.2: Valence region spectra of the untreated TiO_2 substrate and with adsorbed LiI. The difference of both spectra has been taken and multiplied by the factor of five.

According to the difference spectrum, the I 5p orbital is located at a binding energy of $E_B = 3.5$ eV. As for the I 4d emission binding energies a comparison to literature data is given at the end of this chapter (Section 10.3). Due to the much smaller amount of adsorbed iodine from the molten salt, it was not possible to separated the I 5p level by taking a difference spectrum. In order to estimate the binding energy position of the I $5p_{3/2}$ level, we consider the binding energy shift of the I 4d emission level of 0.3 eV as shown by Figure 10.1 right. According to that the I $5p_{3/2}$ level for the I$^-$ of the molten salt is located at 3.2 eV.

10.2 Coadsorption of iodide salt and acetonitrile

On samples containing either LiI or molten salt the same amount of acetonitrile as electrolyte solvent has been coadsorbed. The I 4d emission line ($h\nu = 90$ eV, normalized to beam current) has been measured to detect possible changes induced by the solvent. In order to study the interactions of TiO_2/iodide/solvent, the appropriate photoemission lines, the Ti 2p core level at $h\nu = 600$ eV and the valence band gap region at 90 eV, have been measured. To exclude damping by the redox species and coadsorbed solvent, the Ti 2p emission line has been normalized to the Ti 2p main intensity, and the gap region spectra to the Ti 3p level.

10.2.1 LiI + acetonitrile

I 4d and N 1s core level spectra In Figure 10.3 left, the I 4d level for LiI adsorption and subsequent acetonitrile coadsorption is shown. Interestingly, the emission line does not shift at all upon coadsorbed acetonitrile, although a shift towards higher binding energy for the I 4d and to lower binding energy for the Li 1s is expected due to solvation by surrounding acetonitrile molecules, as shown by gasphase solvation experiments [188]. Possibly the solvation has to be activated in order to show those binding energy shifts described above.

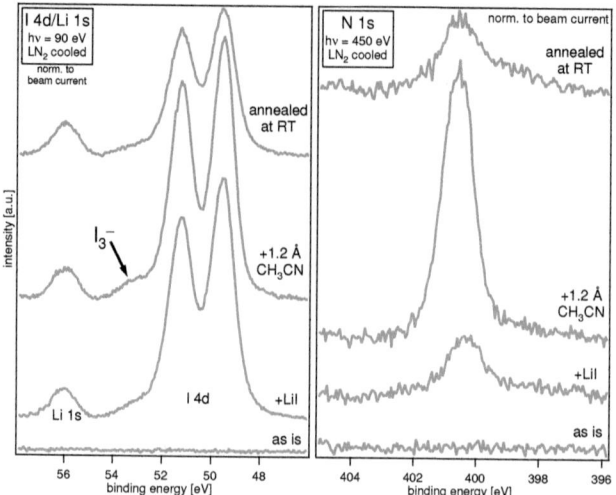

Figure 10.3: left: The I 4d emission line taken at $h\nu = 90$ eV. The spectra are normalized to the beam current. **right:** The N 1s core level, measured with $h\nu = 450$ eV. In the first adsorption step (+LiI) it shows the chemisorption of acetonitrile out from liquid solution.

The N 1s emission line (Fig.10.3 right) shows already upon LiI adsorption an intensity rise. As we carried out the adsorption of LiI from acetonitrile solution at room temperature

10.2 Coadsorption of iodide salt and acetonitrile

(accordingly to Section 4.2.2), it is assumed, that the acetonitrile solvent evaporates. But a fraction of acetonitrile is evidently still left on the surface. Since the LiI adsorption has been carried out at room temperature, acetonitrile is apparently strongly bound (chemisorbed) on the TiO_2 surface. Accordingly, a chemisorption at room temperature of acetonitrile has been observed by other groups with Photoelectron Spectroscopy on different surfaces like Si(001) [301, 302] and Ni(111) [303] surface. Compared to the more noisy spectra of acetonitrile adsorption from the gas phase (Figure 7.1 top right, page 98), a larger amount of acetonitrile molecules remains on the substrate by absorbing LiI from acetonitrile solution.

At slightly higher binding energies, the I 4d level (Fig.10.3 left) shows a small shoulder, which might be due to the formation of I_3^- ions (compare Fig. 11.2 bottom right, page 155). This small contribution increases slightly in intensity upon coadsorption of acetonitrile and decreases upon annealing to room temperature.

Surprisingly the I 4d emission line increases in intensity by coadsorption of the solvent. This is addressed to a more homogeneous distribution of the I^- species over the sample surface. After annealing and evaporation of the coadsorbed solvent to room temperature this distribution seems to be reversed.

Ti 2p and gap region spectra The Ti 2p core level emission has been measured in order to elucidate the substrate/electrolyte interaction (Figure 10.4 top left). The spectra have been normalized against the Ti 2p main emission line to exclude damping effects. One can see, like in the case of all adsorbates so far, a quenching of the gap states upon LiI adsorption. But from the observations of the N 1s spectra it is not clear, whether the quenching of the Ti^{3+} states corresponds to the LiI or the chemisorbed acetonitrile, because acetonitrile alone quenches the Ti^{3+} states as well as we have observed in the adsorption of acetonitrile onto the bare TiO_2 substrate (see Figure 7.3 top left, page 101).

In addition the gap region has been recorded with $h\nu = 90$ eV and normalized to the Ti 3p photoemission line (Fig.10.4 top right). The band gap spectra for the LiI adsorption show clearly a reduction of the V_O states intensity. The intensity changes are very similar to the development of the Ti^{3+} states intensity (Fig.10.4 bottom). Taking the chemisorption of acetonitrile into account (Fig.10.3 right), we cannot conclude whether the LiI or the acetonitrile is responsible for the decrease of the intensity of the V_O states. In addition to this V_O states intensity decrease, we notice an increase of the $Ti3d < E_F$ states upon LiI adsorption.

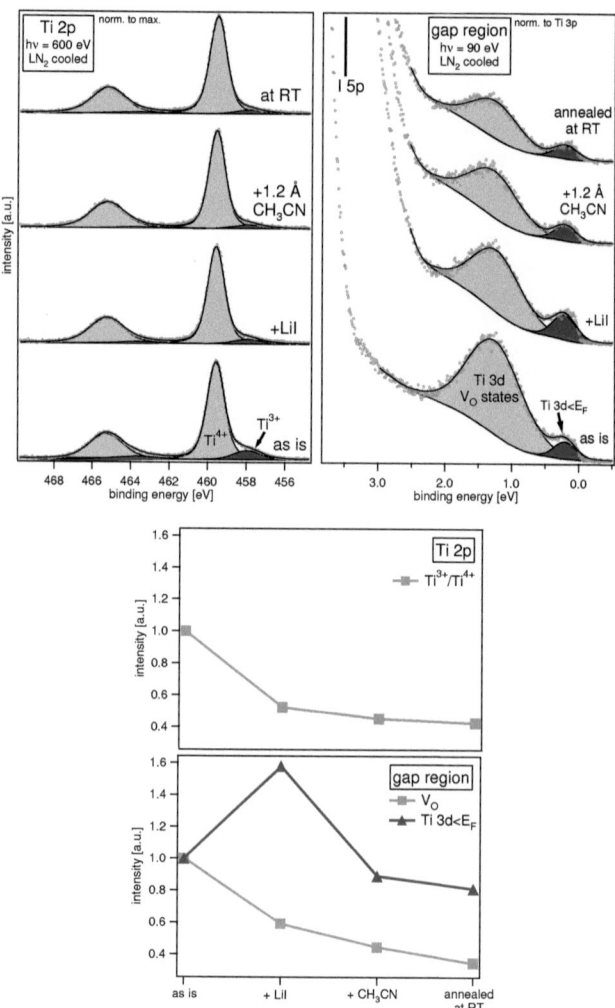

Figure 10.4: top left: The Ti 2p core level recorded at $h\nu = 600$ eV and normalized to the main emission line (Ti^{4+}). **top right:** The gap region ($h\nu = 90$ eV), normalized to the Ti 3p level. **bottom up:** The integrated intensity changes of the Ti^{3+} states relative to the Ti^{4+} states as observed by the Ti 2p emission line. **bottom down:** The intensity changes for the V_O and the $Ti\,3d < E_F$ states, normalized to their respective initial intensity.

10.2.2 molten salt + acetonitrile

I 4d The I 4d emission line with molten salt/acetonitrile coadsorption has been taken at $h\nu = 90$ eV and normalized to the beam current, as displayed by Figure 10.5.

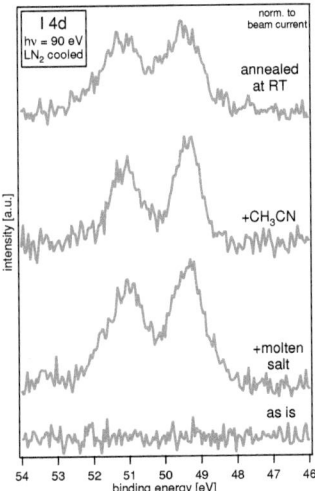

Figure 10.5: The I 4d emission line measured at $h\nu = 90$ eV and normalized to the beam current.

Since the molten salt cation contains two nitrogen atoms (Fig.2.17), the N 1s line is not appropriate to observe a possible chemisorption effect of acetonitrile on the molten salt (the molten salt is like LiI dissolved in acetonitrile). We assume, that by the adsorption of molten salt, some acetonitrile is also adsorbed, as observed for the LiI/acetonitrile coadsorption.

Ti 2p and gap region In line with the observations made for LiI/CH$_3$CN coadsorption the Ti 2p level (10.6 top left) and the gap region spectra (Fig. 10.6 top right) show the same trend. The Ti^{3+} states as well as the V_O states are quenched upon molten salt and acetonitrile coadsorption. As for LiI adsorption, a direct evidence of any interaction between electrolyte salt and acetonitrile is not observed.

In contrast to the increase of the $Ti3d < E_F$ states upon LiI adsorption, here the intensity of the $Ti3d < E_F$ states decreases throughout the coadsorption (Figure 10.6 bottom).

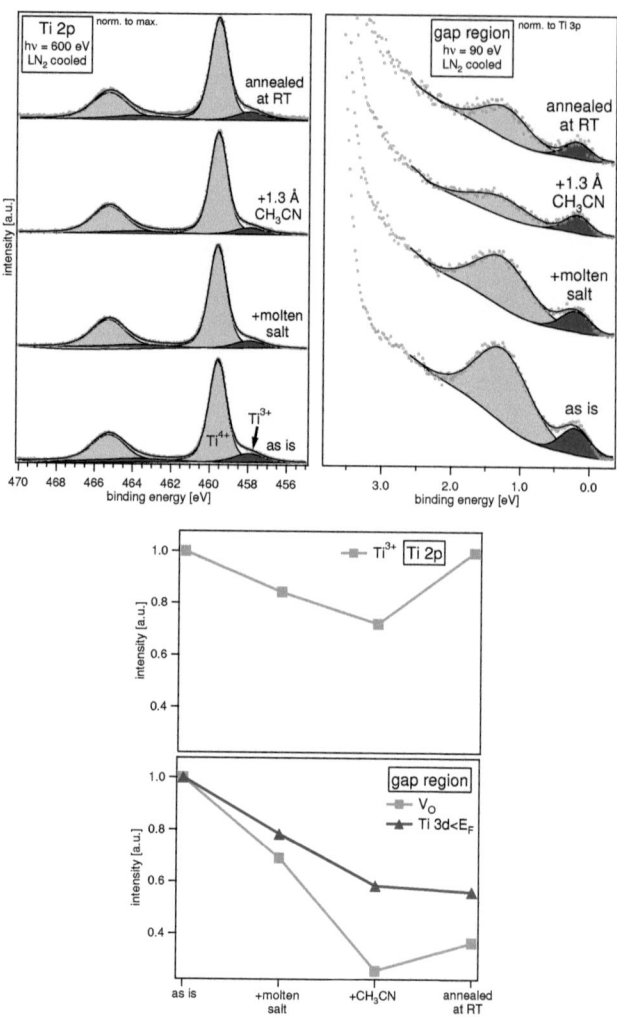

Figure 10.6: top left: The Ti 2p core level ($h\nu = 600$ eV, normalized to main emission line). **top right:** The gap region, taken at $h\nu = 90$ eV; normalized to Ti 3p intensity. **bottom up:** The intensity course of the Ti^{3+} states relative to the Ti^{4+} main emission line. **bottom down:** The intensity course of the V_O and the $Ti\,3d < E_F$ states.

10.3 Discussion of the results

Weber et al. [186] have measured LiI in a water jet by Photoelectron Spectroscopy and referenced their binding energy values against the vacuum level. In order to compare the ionization potential I_p with our binding energy value E_B, the work function ϕ of the TiO$_2$/LiI/acetonitrile interface needs to be subtracted from the ionization potential values of Weber et al. The work function ranges from $\phi = 3.8 - 4.0$ eV, thus we assume 3.9 eV. In Table 10.1 the literature data from Weber et al. is compared to our data (LiI and molten salt respectively).

	Weber et al.	LiI	molten salt
$E_B(I\,4d_{5/2})$ [eV]	49.9	49.6	49.3
$E_B(I\,5p_{3/2})$ [eV]	3.8	3.5	(3.2)
$\Delta E_B(I\,4d_{5/2}\text{-}I\,5p_{3/2})$ [eV]	46.1	46.1	46.1

Table 10.1: The comparison of the binding energy values of Weber et al. [186] with data from this work. The value for the I 5p of the molten salt is in brackets, because it has been deduced from the binding energy difference of the I 4d levels between LiI and molten salt.

The deviation of 0.3 eV (LiI) and 0.6 eV (molten salt) to lower binding energies of our experimental data might be due to screening of the photohole by electron density from the TiO$_2$ substrate. Another reason may be, that a different solvent has been used by Weber et al. (water) than the one used within this work (acetonitrile). According to a photoelectron spectroscopy study of gas phase solvation performed by Markovich et al. the solvation shift of the I 5p orbital by water [304] and acetonitrile [188] differs clearly: Taking a solvent molecule cluster of the same number of molecules (55 solvent molecules), water shifts the I 5p emission about at least 0.5 eV further to higher binding energies than acetonitrile. Taking this into account, our experimentally determined binding energy values match reasonably with the results of Weber et al. [186]. This suggests, that both molten salt and LiI are solvated by surrounding solvent molecules.

In Section 10.1 we observed a binding energy difference of the I 4d emission line of 0.3 eV between the LiI and the molten salt adsorption. In order to explain this shift, we have to consider the differences of both salts, namely the cation. For LiI the Li$^+$ ion is rather small, having an ionic radius of 73 pm, whereas the cation of the molten salt (1-propyl-3-methylimidazolium) is very large with 570 pm. Proportional to the distance between the cation and anion the ionic binding energy decreases. Thus the electrostatic interaction between Li$^+$ and I$^-$ is much larger than for the molten salt cation and I$^-$. The I$^-$ in LiI is more strongly stabilized than in the molten salt, thus the binding energy is shifted to larger values.

11 TiO$_2$ / N3 Ru dye / iodide / solvent interface

In this chapter the complete interface of the working electrode is described by means of the coadsorption of the three components N3 dye, molten salt and acetonitrile.

11.1 Coadsorption of N3 dye, molten salt and acetonitrile

Valence band and gap region spectra After the dye has been deposited onto the TiO$_2$ substrate the molten salt has been coadsorbed. Subsequently three coadsorption steps of increasing acetonitrile coverage have been performed.

Surprisingly after coadsorption of acetonitrile the intensity of the I 4d emission line is increasing (Fig.11.1 left). Obviously, the solvent has a decisive influence on the surface distribution of the I$^-$ ions of the molten salt.

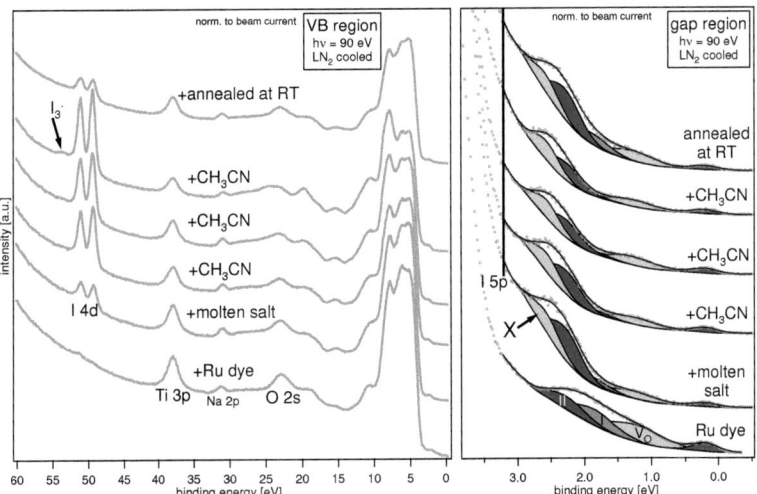

Figure 11.1: left: The valence band, taken at $h\nu = 90$ eV. The intensity has been normalized to the beam current. **right:** The gap region at the same excitation energy $h\nu$, but normalized to the Ti 3p level. For details see text.

The gap region of the coadsorption system is shown by Figure 11.1 right. At the bottom the adsorbed Ru dye shows the two components I and II of the HOMO level at the same binding energies ($E_B = 1.8$ eV (I) and 2.2 eV (II)) as described in Section 8.7, (page 122) by using similar parameters for the fitting procedure. By coadsorbing the molten salt an additional unknown component at $E_B = 2.5$ eV appears (marked with X). This component X is located in its binding energy between the I 5p level at 3.5 eV (Fig. 10.2) and the component II of the HOMO level. During the following coadsorption of acetonitrile, the additional component X is not changed in intensity, whereas the HOMO levels I and II decrease slightly. The assignment of component X is unclear, because no additional component is found in the core level spectra shown below (Fig.11.2 left and middle). However, the component X only shows up by coadsorbing the dye and the molten salt, but by adsorbing the dye or the molten salt alone, it is not present. Regarding possible assignments of these unknown X states, a further discussion is provided in the conclusion section of this chapter (Section 11.2).

S 2p and I 4d core level spectra In Figure 11.2 top left, the S 2p emission line is displayed, taken at $h\nu = 210$ eV. In order to confirm the observations made from the gap region, the S 2p emission of the dye has been deconvoluted into three Voigt profiles according to Figure 8.6 (page 120). Upon molten salt coadsorption, all three S 2p components shift about 70 meV to higher binding energies. After further coadsorption of acetonitrile a similar development as in the case of the acetonitrile coadsorption takes place (Figure 9.2 left, page 129). The binding energy of the components shift by 120 meV and the emission is mainly showing only one species (Figure 11.2 top left). The binding energy shift is in contrast to the results of the acetonitrile coadsorption without I^-, where the binding energies of the components remain at same values.

On the other hand, the I 4d emission line, shown by Figure 11.2 bottom left, shows a shift into the opposite direction: The I 4d emission is located at 100 meV lower binding energies ($E_B = 49.2$ eV) compared to the value for molten salt adsorbed onto the untreated TiO_2 substrate ($E_B = 49.3$ eV). In the discussion section below (Section 11.2), an interpretation of these opposite shifts of the S 2p and the I 4d emission line is given.

In addition, the proceeding coadsorption of acetonitrile shows clearly the rise in intensity of the I 4d line again, although no more molten salt is adsorbed (Figure 11.2 top right). The acetonitrile leads to a shift of 150 meV towards higher binding energies.

Also additional states at higher binding energy ($E_B = 51.7$ eV) appear after coadsorbing acetonitrile, which are addressed as I_3^- states. By a Voigt fitting procedure the binding energy difference to the main line have been determined to be 2.4 eV (Fig.11.2 bottom right).

11.1 Coadsorption of N3 dye, molten salt and acetonitrile

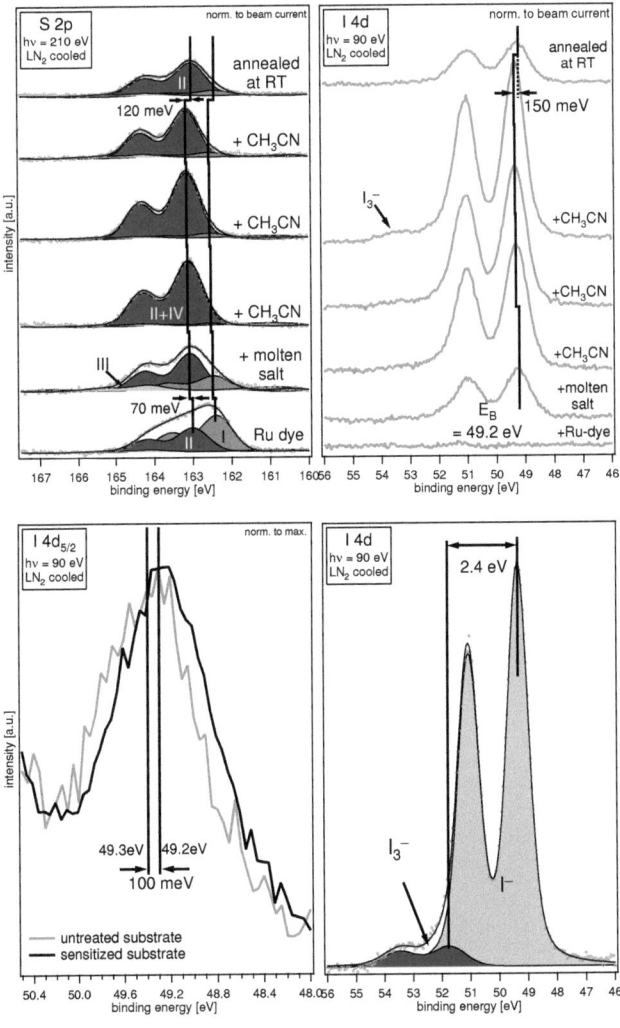

Figure 11.2: top left: The S 2p level, recorded at $h\nu = 210$ eV and normalized to the beam current. **top right:** The I 4d level, measured at $h\nu = 90$ eV and normalized to the beam current. **bottom left:** Comparison between the I $4d_{5/2}$ emission line of the molten salt adsorbed onto the untreated TiO$_2$ substrate (from Figure 10.5 right) and on the dye covered substrate (from top right graph). **bottom right:** I 4d emission line after the third acetonitrile adsorption step (from top right graph) and its deconvolution by a Voigt fitting procedure of the last coadsorption step of acetonitrile on top of Ru dye/LiI showing triiodide

11.2 Discussion of the results

By means of the more homogeneous distribution of the molten salt after acetonitrile adsorption (Fig.11.1 left), we could show, that the molten salt is dissolved in the solvent, probably activated by the synchrotron beam. Thus we could demonstrate, that the model experiments at the TiO_2/dye/molten salt/solvent interface approaches to the real, liquid electrolyte.

An increase of an additional component at higher binding energy in the I 4d spectra course (Fig.11.2) is related to the formation of I_3^-. A possible explanation for the presence of the I_3^- states could be, that I^- ions are oxidized to I_3^- in order to rereduce photooxidized dye.

The additional states in the gap region at $E_B = 2.5$ eV as well as the opposite binding energy shifts of the S 2p and the I 4d emission line upon dye/molten salt coadsorption favors a model of intermolecular interaction. The opposite shifts of S 2p and I 4d (Figure 11.2 top and bottom left) suggest a chemical bonding between iodide and the thiocyanate group. It has been found by Agrell et al. [296, 297] that I_2 and SCN^- interact strongly by forming a I_2SCN^- complex. Consequently new states between the HOMO and the I 5p orbitals are formed. Since their origin is not clear they are called X states in Figure 11.1 right. Since these X states are energetically located between the I 5p and the HOMO level, these X states would be of nonbonding character within this model. This assumption involves the formation of bonding orbitals as well, which are located below the I 5p states. Due to the high DOS of the valence band of the TiO_2 substrate these bonding states cannot be separated clearly. Within this assumption these nonbonding hybrid HOMO-I 5p states may be the appropriate electronic level for the charge transfer of holes from the dye to the I 5p.

However, it has to be clearly stated, that this assignment is only a tentatively hypothesis in order to explain these unknown X states. The shift of the S 2p level towards higher binding energy could also be explained by a solvation of cations of the molten salt, which forms a dipole shell around the sulphur atom, thus leading to the positive binding energy shift by stabilization. Following this approach, the X states might originate from cation surrounded HOMO orbitals.

The assumed electronic alignment of the complete interface is shown by Figure 11.3. All electronic levels, including the V_O and $Ti 3d < E_F$ states, the two HOMO levels, the I 5p states and the attributed HOMO - I 5p hybrid states are included in this picture.

After the photoexcitation of an electron from the (isolated) HOMO level to the LUMO the photohole transfers from the dye molecule to the I^- ion. There are two possible paths possible for the hole transfer: The first (①) is the transfer path, for which it is assumed, that the so attributed hybrid X states are involved in the charge transfer. Following this assumption, the holes need to overcome a barrier of only 0.3 eV in order to leave the dye HOMO level. The second possible hole transfer path (②) does not include the hybrid X states. Assuming this option, the holes need to get over a barrier of even 1.0 eV which is far above thermal energy.

11.2 Discussion of the results

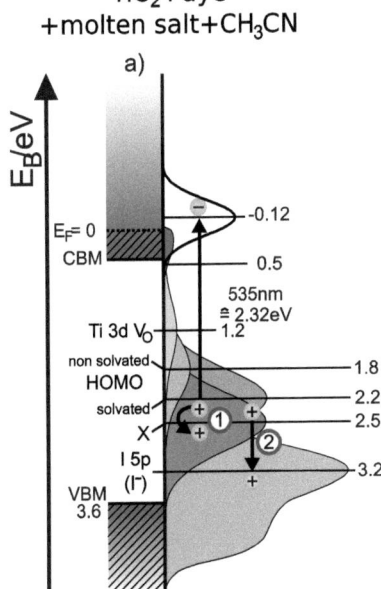

Figure 11.3: The energetic positions of the electronic states of the complete interface of the working electrode in the DSSC as measured in this work. Two possible charge transfer paths for holes from the HOMO to the I 5p states of the I$^-$ ion are illustrated:
①: The X states are involved in the hole transfer.
②: The hole transfer without involving the X states.

It is surprising, that the transfer of the photohole seems to include a more or less large barrier. This is in contradiction to the electrochemical data from literature, where the redox potential of the electrolyte is aligned to the dye HOMO level without barrier (see Figure 1.1, page 15). Therefore it seems safe to conclude that a direct transfer to I$^-$ ions in solution is not an appropriate hole transfer and dye neutralization mechanism after light induced electron injection.

12 Summary & Outlook

12.1 TiO$_2$ substrate

The morphology of nanocrystalline as well as CVD-TiO$_2$ substrate have been characterized. The nanocrystalline TiO$_2$ shows a granular morphology, whereas the CVD-TiO$_2$ varies from granular microstructure to a structure build-up from plates, being arranged perpendicular to the surface. Although the microstructure of both substrates may differ from case to case, it has no significant impact on the measured electronic properties as indicated by a similar gap state distribution.

By GIXRD measurements the crystalline phases of the TiO$_2$ of both nc-TiO$_2$ and CVD-TiO$_2$ have been determined to be anatase. In addition Raman spectroscopy has been applied on the CVD-TiO$_2$, which confirmed the result of the GIXRD measurement. One of the reasons for the use of anatase instead of rutile for Dye Sensitized Solar Cells is the 10 times higher electron mobility in anatase compared to rutile.

In the bandgap region of the substrate, two kinds of gap states have been observed, namely V_O at a binding energy of 1.3 eV, and $Ti3d < E_F$ states, which are located just below the Fermi level. While the V_O states could be assigned as oxygen vacancy surface states, the origin of the $Ti3d < E_F$ states remains unclear. The valence band maximum is located at $E_B = 3.6$ eV, while the band gap of anatase is only 3.2 eV. Assuming the rigid band model, the anatase substrate is a highly degenerated semiconductor with 400 meV of occupied conduction band states.

Also alternative explanations for the origin of the $Ti3d < E_F$ states are possible. Models have been developed, where these $Ti3d < E_F$ states have been either attributed to delocalized states, subsurface states or as states originating from a Ti$_2$O$_3$ surface phase. However, we want to suggest an intermediate case, where the $Ti3d < E_F$ states are located in the subsurface region. The different intensity ratio of V_O and $Ti3d < E_F$ states observed at different photon excitation energies indicates that the $Ti3d < E_F$ states are not located at the outermost surface. On the other hand the partial quenching of the $Ti3d < E_F$ states due to adsorbates is a hint for the surface position of these states.

12.2 Solvent adsorption

Two kinds of solvents, polar acetonitrile and unpolar benzene, have been adsorbed onto the untreated TiO$_2$ substrate. In order to adsorb the solvent from the gas phase onto the sample in the UHV, the sample has been mounted onto a manipulator cooled with liquid nitrogen. The coverage has been estimated by the attenuation of the main Ti 2p substrate signal. The electronic structure of the valence band region of both solvents has been characterized and the respective molecular orbitals have been assigned by comparison to gas phase spectra from literature.

In the case of acetonitrile adsorption a quenching of both Ti^{3+} and V_O states to a similar extent has been observed. To a lower degree, the $Ti3d < E_F$ states are quenched as well. The quenching of electronic states is an indication for an electronic interaction between acetonitrile and the TiO_2 surface. Unlike acetonitrile, benzene does not show such strong interaction with the substrate. Very high coverages are needed to observe a significant decrease in intensity of the Ti^{3+} and V_O states. The $Ti3d < E_F$ states are hardly influenced by the adsorption of benzene.

In the case of acetonitrile, the work function is lowered by 0.35 eV. Since the acetonitrile molecule is a strong dipole, it has been concluded that the dipole is directed towards the substrate surface, i.e. the nitrogen is pointing towards the surface.

Upon annealing to room temperature the solvent evaporates and all quenched states are fully restored in the case of benzene, whereas a fraction of acetonitrile molecules remains on the substrate surface. Therefore it has been deduced, that the adsorption of benzene is completely reversible, while a fraction of acetonitrile molecules is obviously chemisorbed onto the TiO_2 surface.

Since the surface defect states are quenched by acetonitrile, it has been concluded that acetonitrile is not only a medium for dissolving the electrolyte salt. Acetonitrile also reduces a possible recombination pathway of photoinjected electrons via the detrimental surface defect states.

12.3 Dye adsorption

Dye adsorption has been performed from ethanol solution at room temperature and normal pressure argon atmosphere. In order to yield monolayer coverage the sample has been rinsed with ethanol removing nonbonded dye molecules. The respective atomic components within the core level spectra have been deconvoluted by Voigt profiles. The integrated intensities of the respective components match reasonably with the expected number of elements given by the molecular structure.

Regarding the interaction of the dye molecules with the TiO_2 surface earlier Photoelectron Spectroscopy experiments, performed by Westermark et al. [21], have been confirmed, that the V_O states are quenched upon dye adsorption. In line with the V_O state quenching, an intensity decrease of the Ti^{3+} states has been observed. Thus the dye molecules are bound to the same sites as the solvent molecules. But the degree of quenched adsorption sites is smaller in the case of dye adsorption compared to solvent adsorption. It has been deduced, that the much larger dye molecules are not able to reach such a large amount of adsorption sites per area like the much smaller solvent molecules.

12.4 Dye/solvent coadsorption

The coadsorption of either polar acetonitrile or unpolar benzene molecules onto the dye covered substrate surface has been accomplished. In the case of acetonitrile coadsorption, the intensity changes of the appropriate core levels and the HOMO clearly indicates a reorientation and solvation of the dye molecules. In line the work function changes are explained in terms of a dipole model, which indicates a rearrangement of the dye with respect to the substrate. In contrast, the coadsorption of unpolar benzene molecules does not involve an orientational change of the dye molecules. Thus it is concluded that the reorientation depends on the polarity of the coadsorbed solvent.

These results are of high relevance for the function of the DSSC device. The directional transfer of holes and electrons is favored by the distinct geometry of the dye molecule, often referred to as *vectorial charge transfer* [1]. This implies, that the dye ideally works as an electronic membrane, transferring holes and electrons in opposite directions. Therefore the orientation of the dye coupling onto the TiO_2 substrate is the key point to allow an effective charge separation.

Moreover, a model of the alignment of the electronic states, either belonging to the substrate, the dye, or the coadsorbed solvent molecules, has been developed. In order to estimate the binding energy of the LUMO of the dye molecule ($E_B = -0.12$ eV), the maximum of optical absorption of the dye (535 nm $\hat{=}$ 2.32 eV) has been compared to the SXPS data in this work. The model of the electronic alignment at the TiO_2/dye/solvent interface suggests, that photoexcited electrons in the LUMO level are injected into the TiO_2 semiconductor, driven by a potential drop of 0.42 eV.

12.5 Electrolyte salt/solvent coadsorption

Two electrolyte salts, LiI and the molten salt PMII (1-propyl-3-methylimidazolium iodide) have been adsorbed onto the untreated TiO_2 substrate. The adsorption has been performed from acetonitrile solution at room temperature and in argon atmosphere. These are the same ambient conditions like for the dye adsorption. The comparison between the two salts revealed a binding energy difference of 300 meV between the respective I 4d photoemission lines (LiI: $E_B = 49.6$ eV; molten salt: 49.3 eV). With these values the energy position of the I 5p states for the molten salt (3.2 eV) are deduced at 300 meV lower binding energy compared to the I 5p states belonging to LiI (3.5 eV).

One shortcoming of adsorbing the electrolyte salt from solution is the fact that only one redox species, the reduced I^- ion, have been adsorbed. Since iodine (I_2) sublimes, it is not possible to adsorb the oxidized redox species, the I_3^- ion, from solution. Since the focus of this work is on the working electrode half cell, only the reduced redox species is of prime interest. Moreover, by coadsorbing the I_3^- species, a ligand exchange reaction of the thiocyanate groups is promoted, which has been observed by other groups [296, 297].

In the case of LiI the I 4d emission line shows a shoulder at 2.4 eV higher binding energies, which is increasing in intensity with proceeding coadsorption of acetonitrile. Thus an attribution of this shoulder to the I_3^- species has been suggested.

12.6 Dye/electrolyte salt/solvent coadsorption

The coadsorption of the electrolyte, including an I^- anion, the cation (1-propyl-3-methylimidazolium), and the solvent acetonitrile has been performed onto the dye sensitized substrate surface.

Additional valence states, referred to as X states, closely below the HOMO level are formed. The origin of these states is yet unclear, but it has been suggested, that they arise from a chemical bonding of the I^- ions with the thiocyanate group. This assumption is justified by the fact, that this X states only appear, if both dye *and* electrolyte salt are coadsorbed. Following this assumption, molecular orbitals are formed. Since these states are located between the I 5p states and the HOMO level, these states are addressed as nonbonding molecular orbitals between the HOMO level and the I 5p states. Within this suggested model the possible effect of these states regarding the device function has been elucidated. If these X states are involved in the hole transfer from the HOMO to the I^- redox couple species, an energy barrier of 0.3 eV for the hole injection into the electrolyte has been estimated. If the X states are not involved in the hole transfer, then the barrier or the hole injection is with 1.0 eV much too large.

The more homogeneous distribution of the I^- induced by the coadsorbed solvent suggests that the electrolyte salt is dissolved in the solvent, probably activated by the synchrotron beam. Thus it has been demonstrated, that the model experiments at the TiO_2/dye/electrolyte salt/solvent interface correspond reasonably to the real, liquid electrolyte.

12.7 Outlook

In this work, a substrate material, which is widely applied in the DSSC device, has been used. In order to investigate the interactions of the diverse adsorbates with the substrate of defined anatase surfaces, anatase single crystals need to be prepared.

To confirm the reorientation of the dye molecules on the substrate surface, dye molecules need to be adsorbed on a flat surface of a defined single crystalline anatase sample. Investigating acetonitrile and benzene coadsorption onto the dye covered single crystal are expected to yield contrary results. In order to yield clear results, e.g. by analyzing the coadsorption with X-Ray Absorption Spectroscopy, a suitable dye has to be chosen with the π electron systems in one plane, which should show an angular dependency of absorption intensity on the polarization of the synchrotron light.

In order to prepare the complete electrolyte in UHV, the oxidized redox species, the I_3^- ion, needs also to be adsorbed. Maybe the origin of the additional component, which only appears in the presence of dye, solvent and electrolyte, can be clarified by performing experiments with the complete electrolyte. This can only be achieved by freezing the electrolyte while still liquid. It has to be taken great care in order to reduce possible contaminations by water adsorption.

4-tert butylpyridine is a widely used additive for improving the performance of the DSSC. It is observed, that its application causes an increase of the photovoltage due to moving the conduction band edge upward [76, 305]. The electron lifetime is increased, which might be because of a decreased accessibility of I_3^- ions to the TiO_2 surface, thus preventing recombination. To elucidate the mode of action of this additive, a combined study of electrical measurements and UPS has been accomplished by Dürr et al. [306]. They found a decrease of band gap states due to adsorption of 4-TBP, therefore concluding a decrease of related recombination centers. It would also be of interest to investigate the influence of 4-tert butylpyridine with respect to the dye molecules and the electrolyte by coadsorption experiments.

Bibliography

[1] O'Regan, B., Moser, J., Anderson, M., and Grätzel, M. *Journal of Physical Chemistry* **94**(24), 8720–8726 (1990).

[2] Becquerel, E. *Comptes Rendus de l'Académie des Sciences* **9**, 145–149 (1839).

[3] West, W. *Photographic Science and Engineering* **18**(1), 35–48 (1974).

[4] Moser, J. *Monatshefte für Chemie / Chemical Monthly* **8**(1), 373–373 (1887).

[5] Nelson, R. C. *The Journal of Physical Chemistry* **69**(3), 714–718 (1965).

[6] O'Regan, B. and Grätzel, M. a. *Nature* **353**, 737–740 (1991).

[7] Grätzel, M. *Nature* **414**, 338–344 (2001).

[8] Moser, J., Bonnote, P., and Grätzel, M. *Coordination Chemistry Reviews* **171**, 245–250 (1998).

[9] Hagfeldt, A. and Grätzel, M. *Accounts of Chemical Research* **33**(5), 269–277 (2000).

[10] Grätzel, M. *Journal of Photochemistry and Photobiology C* **4**, 145–153 (2003).

[11] Cahen, D., Hodes, G., Grätzel, M., Guillemoles, J. F., and Riess, I. *Journal of Physical Chemistry B* **104**(9), 2053–2059 (2000).

[12] Hannappel, T., Burfeindt, B., Storck, W., and Willig, F. *Journal of Physical Chemistry B* **101**(35), 6799–6802 (1997).

[13] Anderson, N. A. and Lian, T. *Coordination Chemistry Reviews* **248**(13-14), 1231–1246 (2004).

[14] Papageorgiou, N. *Coordination Chemistry Reviews* **248**(13-14), 1421–1446 (2004).

[15] Hara, K. and Arakawa, H. *Dye-sensitizes Solar Cells*, chapter 15, 663–700. John Wiley & Sons (2003).

[16] Hagfeldt, A. and Grätzel, M. *Chemical Reviews* **95**(1), 49–68 (1995).

[17] Gregg, B. A. *Coordination Chemistry Reviews* **248**(13-14), 1215–1224 (2004).

[18] Nazeeruddin, M. K., Kay, A., Rodicio, I., Humpbry-Baker, R., Müller, E., Liska, P., Vlachopoulos, N., and Grätzel, M. *Journal of the American Chemical Society* **115**, 6382–6390 (1993).

[19] Hill, I. G., Kahn, A., Soos, Z. G., and Pascal, R. A. *Chemical Physics Letters* **327**(3-4), 181–188 (2000).

[20] Snook, J. H., Samuelson, L. A., Kumar, J., Kim, Y. G., and Whitten, J. E. *Organic Electronics* **6**(2), 55–64 (2005).

[21] Westermark, K. *Chemical Physics* **285**(1), 157–165 (2002).

[22] Liu, G., Jaegermann, W., He, J., Sundström, V., and Sun, L. *Journal of Physical Chemistry B* **106**, 5814–5819 (2002).

[23] Henrich, V. and Cox, P. *The Surface Science of Metal Oxides*. Cambridge University Press, (1994).

[24] Helmholtz, H. *Annalen der Physik und Chemie* **243**, 337–382 (1879).

[25] Atkins, P. *Physical Chemistry*. Oxford University Press, (1986).

[26] Wedler, G. *Lehrbuch der Physikalischen Chemie*. VCH Verlagsgesellschaft, (1987).

[27] Kron, G. *Ladungsträgertransport in farbstoffsensibilisierten Solarzellen auf Basis von nanoporösem TiO2*. PhD thesis, Universität Stuttgart, (2003).

[28] Gouy, G. *Journal of Physics* **9**(4), 457–468 (1910).

[29] Chapman, D. *Philosophical Magazine* **25**, 475–481 (1913).

[30] Stern, O. *Zeitschrift für Elektrochemie* **30**, 508 (1924).

[31] Zaban, A., Ferrere, S., and Gregg, B. *Journal of Physical Chemistry B* **102**(2), 452–460 (1998).

[32] Gregg, B., Zaban, A., and Ferrere, S. *Zeitschrift für Physikalische Chemie* **212**, 11–22 (1999).

[33] Nozik, A. *Annual Review of Physical Chemistry* **29**, 189–222 (1978).

[34] Gerischer, H. *Electrochimica Acta* **34**(8), 1005–1009 (1989).

[35] Yan, S. and Hupp, J. *Journal of Physical Chemistry* **100**(17), 6867–6870 (1996).

[36] Tachibana, Y., Moser, J. E., Gratzel, M., Klug, D. R., and Durrant, J. R. *Journal of Physical Chemistry* **100**(51), 20056–20062 (1996).

[37] Kelly, C. A., Farzad, F., Thompson, D. W., Stipkala, J. M., and Meyer, G. J. *Langmuir* **15**(20), 7047–7054 (1999).

[38] Haque, S. A., Palomares, E., Cho, B. M., Green, A. N. M., Hirata, N., Klug, D. R., and Durrant, J. R. *Journal of the American Chemical Society* **127**(10), 3456–3462 (2005).

[39] Furube, A., Murai, M., Watanabe, S., Hara, K., Katoh, R., and Tachiya, M. *Journal of Photochemistry and Photobiology A -Chemistry* **182**(3), 273–279 (2006).

[40] Nelson, J. *Charge Transport in Dye-sensitized Systems*, chapter 5.3, 432–474. Wiley-VCH (2003).

[41] Montanari, I., Nelson, J., and Durrant, J. *Journal of Physical Chemistry B* **106**(47), 12203–12210 (2002).

[42] Fillinger, A. and Parkinson, B. *Journal of the Electrochemical Society* **146**(12), 4559–4564 (1999).

[43] Peter, L. M. and Wijayantha, K. G. U. *Electrochimica Acta* **45**(28), 4543–4551 (2000).

[44] Vanmaekelbergh, D. and de Jongh, P. *Physical Review B* **61**(7), 4699–4704 (2000).

[45] Cao, F., Oskam, G., Meyer, G., and Searson, P. *Journal of Physical Chemistry* **100**(42), 17021–17027 (1996).

[46] Kambili, A., Walker, A. B., Qiu, F. L., Fisher, A. C., Savin, A. D., and Peter, L. M. *Physica E: Low-dimensional Systems and Nanostructures* **14**(1-2), 203–209 (2002).

[47] Solbrand, A., Lindstrom, H., Rensmo, H., Hagfeldt, A., Lindquist, S. E., and Sodergren, S. *Journal of Physical Chemistry B* **101**(14), 2514–2518 (1997).

[48] Frank, A. J., Kopidakis, N., and van de Lagemaat, J. *Coordination Chemistry Reviews* **248**(13-14), 1165–1179 (2004).

[49] Bisquert, J. *Physical Review Letters* **91**(1), 010602/1–010602/4 (2003).

[50] Cass, M., Qiu, F., Walker, A., Fisher, A., and Peter, L. *Journal of Physical Chemistry B* **107**(1), 113–119 (2003).

[51] Kambe, S., Nakade, S., Kitamura, T., Wada, Y., and Yanagida, S. *Journal of Physical Chemistry B* **106**(11), 2967–2972 (2002).

[52] van de Lagemaat, J. and Frank, A. *Journal of Physical Chemistry B* **105**(45), 11194–11205 (2001).

[53] Nelson, J. and Chandler, R. *Coordination Chemistry Reviews* **248**(13-14), 1181–1194 (2004).

[54] Peter, L. *Journal of Electroanalytical Chemistry* **599**(2), 233–240 (2007).

[55] Jaegermann, W. In *Modern Aspects of Electrochemistry*, White, R; Conway, B., editor, volume 30 of *Modern Aspects of Electrochemistry*, chapter 1, 1–185. Plenum Press, New York (1996).

[56] Albery, W. and Bartlett, P. *Journal of the Electrochemical Society* **131**(2), 315–325 (1984).

[57] Bisquert, J., Garcia-Belmonte, G., and Fabregat-Santiago, F. *Journal of Solid State Electrochemistry* **3**(6), 337–347 (1999).

[58] Gregg, B. *The Essential Interface: Studies in Dye Sensitizes Solar Cells*, chapter 2, 51–89. Marcel Dekker (2003).

[59] van de Lagemaat, J., Park, N. G., and Frank, A. J. *Journal of Physical Chemistry B* **104**(9), 2044–2052 (2000).

[60] Solbrand, A., Henningsson, A., Sodergren, S., Lindstrom, H., Hagfeldt, A., and Lindquist, S. *Journal of Physical Chemistry B* **103**(7), 1078–1083 (1999).

[61] Kopidakis, N., Schiff, E., Park, N., van de Lagemaat, J., and Frank, A. *Journal of Physical Chemistry B* **104**(16), 3930–3936 (2000).

[62] Nister, D., Keis, K., Lindquist, S., and Hagfeldt, A. *Solar Energy Materials and Solar Cells* **73**(4), 411–423 (2002).

[63] van Roosbroeck, W. *Physical Review* **91**(2), 282–289 (1953).

[64] Benkstein, K., Kopidakis, N., van de Lagemaat, J., and Frank, A. *Journal of Physical Chemistry B* **107**(31), 7759–7767 (2003).

[65] Stauffer, D. and Aharony, A. *Introduction to Percolation Theory*. CRC Press, (1994).

[66] Schwarzburg, K. and Willig, F. *Applied Physics Letters* **58**(22), 2520–2522 (1991).

[67] de Jongh, P. and Vanmaekelbergh, D. *Physical Review Letters* **77**(16), 3427–3430 (1996).

[68] Fisher, A., Peter, L., Ponomarev, E., Walker, A., and Wijayantha, K. *Journal of Physical Chemistry B* **104**(5), 949–958 (2000).

[69] van de Lagemaat, A. J. F. *Journal of Physical Chemistry B* **104**, 4292–4294 (2000).

[70] Nelson, J. *Physical Review B* **59**(23), 15374–15380 (1999).

[71] Bisquert, J. and Vikhrenko, V. *Journal of Physical Chemistry B* **108**(7), 2313–2322 (2004).

[72] Vanmaekelbergh, D. and de Jongh, P. *Journal of Physical Chemistry B* **103**(5), 747–750 (1999).

[73] Duffy, N., Peter, L., Rajapakse, R., and Wijayantha, K. *Electrochemistry Communications* **2**(9), 658–662 (2000).

[74] Bailes, M., Cameron, P., Lobato, K., and Peter, L. *Journal of Physical Chemistry B* **109**(32), 15429–15435 (2005).

[75] Kopidakis, N., Neale, N., Zhu, K., van de Lagemaat, J., and Frank, A. *Applied Physics Letters* **87**(20) (2005).

[76] Schlichthörl, G., Huang, S., Sprague, J., and Frank, A. *Journal of Physical Chemistry B* **101**(41), 8141–8155 (1997).

[77] Kopidakis, N., Benkstein, K., van de Lagemaat, J., and Frank, A. *Journal of Physical Chemistry B* **107**(41), 11307–11315 (2003).

[78] Zhu, K., Kopidakis, N., Neale, N. R., van de Lagemaat, J., and Frank, A. J. *Journal of Physical Chemistry B* **110**(50), 25174–25180 (2006).

[79] Ferrere, S. and Gregg, B. *Journal of Physical Chemistry B* **105**(32), 7602–7605 (2001).

[80] Gregg, B. A., Chen, S. G., and Ferrere, S. *Journal of Physical Chemistry B* **107**(13), 3019–3029 (2003).

[81] Durrant, J. R., Haque, S. A., and Palomares, E. *Coordination Chemistry Reviews* **248**(13-14), 1247–1257 (2004).

[82] Huang, S., Schlichthorl, G., Nozik, A., Gratzel, M., and Frank, A. *Journal of Physical Chemistry B* **101**(14), 2576–2582 (1997).

[83] Duffy, N., Peter, L., Rajapakse, R., and Wijayantha, K. *Journal of Physical Chemistry B* **104**(38), 8916–8919 (2000).

[84] Haque, S., Tachibana, Y., Klug, D., and Durrant, J. *Journal of Physical Chemistry B* **102**(10), 1745–1749 (1998).

[85] Haque, S. A., Tachibana, Y., Willis, R. L., Moser, J. E., Gratzel, M., Klug, D. R., and Durrant, J. R. *Journal of Physical Chemistry B* **104**(3), 538–547 (2000).

[86] Gregg, B., Pichot, F., Ferrere, S., and Fields, C. *Journal of Physical Chemistry B* **105**(7), 1422–1429 (2001).

Bibliography

[87] Bach, U., Lupo, D., Comte, P., Moser, J., Weissörtel, F., Salbeck, J., Spreitzer, H., and Grätzel, M. *Nature* **395**, 583–585 (1998).

[88] Bach, U. *Solid State Dye-sensitized Solar Cells - An Alternative Route Towards Low-Cost Photovoltaic Devices*, chapter 5.4, 475–491. Wiley-VCH (2003).

[89] Krüger, J., Plass, R., Cevey, L., Piccirelli, M., Gratzel, M., and Bach, U. *Applied Physics Letters* **79**(13), 2085–2087 (2001).

[90] Krüger, J., Plass, R., Gratzel, M., and Matthieu, H. *Applied Physics Letters* **81**(2), 367–369 (2002).

[91] Dai, Q., MacFarlane, D., and Forsyth, M. *Solid State Ionics* **177**(3-4), 395–401 (2006).

[92] Kubo, W., Kambe, S., Nakade, S., Kitamura, T., Hanabusa, K., Wada, Y., and Yanagida, S. *Journal of Physical Chemistry B* **107**(18), 4374–4381 (2003).

[93] Stathatos, E., Lianos, P., Lavrencic-Stangar, U., and Orel, B. *Advanced Materials* **14**(5), 354–357 (2002).

[94] O'Regan, B., Lenzmann, F., Muis, R., and Wienke, J. *Chemistry of Materials* **14**(12), 5023–5029 (2002).

[95] Palomares, E., Clifford, J., Haque, S., Lutz, T., and Durrant, J. *Chemical Communications* (14), 1464–1465 (2002).

[96] Chappel, S., Grinis, L., Ofir, A., and Zaban, A. *Journal of Physical Chemistry B* **109**(5), 1643–1647 (2005).

[97] Tributsch, H. *Coordination Chemistry Reviews* **248**(13-14), 1511–1530 (2004).

[98] Diebold, U. *Surface Science Reports* **48**, 53–229 (2003).

[99] Vegard, L. *Philosophical Magazine* **32**, 505–518 (1916).

[100] Mo, S. D. and Ching, W. Y. *Physical Review B* **51**(19), 13023–13032 (1995).

[101] Kavan, L., Gratzel, M., Gilbert, S. E., Klemenz, C., and Scheel, H. J. *Journal of the American Chemical Society* **118**(28), 6716–6723 (1996).

[102] Sodergren, S., Siegbahn, H., Rensmo, H., Lindstrom, H., Hagfeldt, A., and Lindquist, S. E. *Journal of Physical Chemistry B* **101**(16), 3087–3090 (1997).

[103] Hengerer, R., Kavan, L., Krtil, P., and Grätzel, M. *Journal of the Electrochemical Society* **147**(4), 1467–1472 (2000).

[104] Tang, H., Berger, H., Schmid, P., Levy, F., and Burri, G. *Solid State Communications* **87**(9), 847–850 (1993).

[105] Pascual, J., Camassel, J., and Mathieu, H. *Physical Review B* **18**(10), 5606–5614 (1978).

[106] Wells, A. *Structural Inorganic Chemistry*. Oxford University Press, (1975).

[107] Lazzeri, M., Vittadini, A., and Selloni, A. *Physical Review B* **63**(15), 155409/1–155409/9 (2001).

[108] Ramamoorthy, M., Vanderbilt, D., and Kingsmith, R. *Physical Review B* **49**(23), 16721–16727 (1994).

[109] Diebold, U., Ruzycki, N., Herman, G. S., and Selloni, A. *Catalysis Today* **85**(2-4), 93–100 (2003).

[110] Ruzycki, N., Herman, G., Boatner, L., and Diebold, U. *Surface Science* **529**(1-2), L239–L244 (2003).

[111] Liang, Y., Gan, S. P., Chambers, S. A., and Altman, E. I. *Physical Review B* **63**(23), 235402/1–235402/7 (2001).

[112] Barbe, C., Arendse, F., Comte, P., Jirousek, M., Lenzmann, F., Shklover, V., and Gratzel, M. *Journal of the American Ceramic Society* **80**(12), 3157–3171 (1997).

[113] http://www.mindat.org/min-3486.html.

[114] Zhang, H. and Banfield, J. *Journal of Materials Chemistry* **8**(9), 2073–2076 (1998).

[115] Barnard, A. and Zapol, P. *Journal of Physical Chemistry B* **108**(48), 18435–18440 (2004).

[116] Barnard, A. and Zapol, P. *Physical Review B* **70**(23), 235403/1–235403/13 (2004).

[117] Naicker, P., Cummings, P., Zhang, H., and Banfield, J. *Journal of Physical Chemistry B* **109**(32), 15243–15249 (2005).

[118] Ranade, M., Navrotsky, A., Zhang, H., Banfield, J., Elder, S., Zaban, A., Borse, P., Kulkarni, S., Doran, G., and Whitfield, H. *Proceedings of the National Academy of Sciences of the United States of America* **99**, 6476–6481 (2002).

[119] Navrotsky, A. *Geochemical Transactions* **4**(6), 34–37 (2003).

[120] Levchenko, A. A., Li, G., Boerio-Goates, J., Woodfield, B. F., and Navrotsky, A. *Chemistry of Materials* **18**(26), 6324–6332 (2006).

[121] Navrotsky, A. *Journal of Chemical Thermodynamics* **39**(1), 1–9 (2007).

[122] Gong, X.-Q., Selloni, A., Batzill, M., and Diebold, U. *Nature Materials* **5**(8), 665–670 (2006).

[123] Hebenstreit, W., Ruzycki, N., Herman, G., Gao, Y., and Diebold, U. *Physical Review B* **62**(24), R16334–R16336 (2000).

[124] Vittadini, A., Selloni, A., Rotzinger, F., and Grätzel, M. *Physical Review Letters* **81**(14), 2954–2957 (1998).

[125] Kim, J., Kang, M., Kim, Y., Won, J., Park, N., and Kang, Y. *Chemical Communications* (14), 1662–1663 (2004).

[126] Tanaka, Y. and Suganuma, M. *Journal of Sol-Gel Science and Technology* **22**(1-2), 83–89 (2001).

[127] Chen, Y.-F., Lee, C.-Y., Yeng, M.-Y., and Chiu, H.-T. *Journal of Crystal Growth* **247**(3-4), 363–370 (2003).

[128] Li, M., Hebenstreit, W., Diebold, U., Tyryshkin, A., Bowman, M., Dunham, G., and Henderson, M. *Journal of Physical Chemistry B* **104**(20), 4944–4950 (2000).

[129] Huber, B., Brodyanski, A., Scheib, M., Orendorz, A., Ziegler, C., and Gnaser, H. *Thin Solid Films* **472**(1-2), 114–124 (2005).

[130] Asahi, R., Taga, Y., Mannstadt, W., and Freeman, A. J. *Physical Review B* **61**(11), 7459–7465 (2000).

[131] Thomas, A. G., Flavell, W. R., Kumarasinghe, A. R., Mallick, A. K., Tsoutsou, D., Smith, G. C., Stockbauer, R., Patel, S., Grätzel, M., and Hengerer, R. *Physical Review B* **67**(3), 035110/1–035110/7 (2003).

[132] Zhang, Z. M., Jeng, S. P., and Henrich, V. E. *Physical Review B* **43**(14), 12004–12011 (1991).

[133] Wang, Y. and Doren, D. *Solid State Communications* **136**(3), 142–146 (2005).

[134] Thomas, A. G., Flavell, W. R., Mallick, A. K., Kumarasinghe, A. R., Tsoutsou, D., Khan, N., Chatwin, C., Rayner, S., Smith, G. C., Stockbauer, R. L., Warren, S., Johal, T. K., Patel, S., Holland, D., Taleb, A., and Wiame, F. *Physical Review B* **75**(3), 035105/12–035105/12 (2007).

[135] Heise, R., Couths, R., and Witzel, S. *Solid State Communications* **84**(6), 599–602 (1992).

[136] Sorantin, P. and Schwarz, K. *Inorganic Chemistry* **31**(4), 567–576 (1992).

[137] Diebold, U., Lehman, J., Mahmoud, T., Kuhn, M., Leonardelli, G., Hebenstreit, W., Schmid, M., and Varga, P. *Surface Science* **411**(1-2), 137–153 (1998).

[138] Hengerer, R., Bolliger, B., Erbudak, M., and Grätzel, M. *Surface Science* **460**(1-3), 162–169 (2000).

[139] Takeuchi, M., Onozaki, Y., Matsumura, Y., Uchida, H., and Kuji, T. *Nuclear Intruments and Methods in Physics Research Section B* **206**, 259–263 (2003).

[140] Orendorz, A., Wusten, J., Ziegler, C., and Gnaser, H. *Applied Surface Science* **252**(1), 85–88 (2005).

[141] Göpel, W., Anderson, J., Frankel, D., Jaehnig, M., Phillips, K., Schafer, J., and Rocker, G. *Surface Science* **139**(2-3), 333–346 (1984).

[142] Sanjines, R., Tang, H., Berger, H., Gozzo, F., Margaritondo, G., and Levy, F. *Journal of Applied Physics* **75**(6), 2945–2951 (1994).

[143] Westermark, K., Rensmo, H., Lees, A. C., Vos, J. G., and Siegbahn, H. *Journal of Physical Chemistry B* **106**(39), 10108–10113 (2002).

[144] Huber, R., Sporlein, S., Moser, J., Gratzel, M., and Wachtveitl, J. *Journal of Physical Chemistry B* **104**(38), 8995–9003 (2000).

[145] Bisquert, J., Zaban, A., and Salvador, P. *Journal of Physical Chemistry B* **106**(34), 8774–8782 (2002).

[146] Nazeeruddin, M. and Grätzel, M. *Dyes for Semiconductor Sensitization*, chapter 5.2, 407–431. Wiley-VCH (2003).

[147] Polo, A., Itokazu, M., and Iha, N. *Coordination Chemistry Reviews* **248**(13-14), 1343–1361 (2004).

[148] Kalyanasundaram, K. and Grätzel, M. *Coordination Chemistry Reviews* **177**, 347–414 (1998).

[149] Shklover, V., Ovchinnikov, Y. E., Braginsky, L. S., Zakeeruddin, S. M., and Grätzel, M. *Chemistry of Materials* **10**(9), 2533–2541 (1998).

[150] Rensmo, H., Lunell, S., and Siegbahn, H. *Journal of Photochemistry and Photobiology A - Chemistry* **114**(2), 117–124 (1998).

[151] Srikanth, K., Marathe, V., and Mishra, M. *International Journal of Quantum Chemistry* **89**(6), 535–549 (2002).

[152] Monat, J. E., Rodriguez, J. H., and McCusker, J. K. *Journal of Physical Chemistry A* **106**(32), 7399–7406 (2002).

[153] Guillemoles, J., Barone, V., Joubert, L., and Adamo, C. *Journal of Physical Chemistry A* **106**(46), 11354–11360 (2002).

[154] Zabri, H., Gillaizeau, I., Bignozzi, C., Caramori, S., Charlot, M., Cano-Boquera, J., and Odobel, F. *Inorganic Chemistry* **42**(21), 6655–6666 (2003).

[155] De Angelis, F., Fantacci, S., and Selloni, A. *Chemical Physics Letters* **389**(1-3), 204–208 (2004).

[156] Nazeeruddin, M., De Angelis, F., Fantacci, S., Selloni, A., Viscardi, G., Liska, P., Ito, S., Takeru, B., and Grätzel, M. *Journal of the American Chemical Society* **127**(48), 16835–16847 (2005).

[157] De Angelis, F., Tilocca, A., and Selloni, A. *Journal of the American Chemical Society* **126**(46), 15024–15025 (2004).

[158] Nazeeruddin, M., Zakeeruddin, S., Humphry-Baker, R., Jirousek, M., Liska, P., Vlachopoulos, N., Shklover, V., Fischer, C., and Grätzel, M. *Inorganic Chemistry* **38**(26), 6298–6305 (1999).

[159] Johansson, E. M. J., Hedlund, M., Siegbahn, H., and Rensmo, H. *Journal of Physical Chemistry B* **109**(47), 22256–22263 (2005).

[160] Lu, Y., Choi, D., Nelson, J., Yang, O., and Parkinson, B. *Journal of the Electrochemical Society* **153**(8), E131–E137 (2006).

[161] Finnie, K., Bartlett, J., and Woolfrey, J. *Langmuir* **14**(10), 2744–2749 (1998).

[162] Persson, P. and Lundqvist, M. *Journal of Physical Chemistry B* **109**(24), 11918–11924 (2005).

[163] Zubavichus, Y., Slovokhotov, Y., Nazeeruddin, M., Zakeeruddin, S., Grätzel, M., and Shklover, V. *Chemistry of Materials* **14**(8), 3556–3563 (2002).

[164] Duffy, N., Dobson, K., Gordon, K., Robinson, B., and McQuillan, A. *Chemical Physics Letters* **266**(5-6), 451–455 (1997).

[165] Kilså, K., Mayo, E. I., Brunschwig, B. S., Gray, H. B., Lewis, N. S., and Winkler, J. R. *Journal of Physical Chemistry B* **108**(40), 15640–15651 (2004).

[166] Onishi, H., Aruga, T., and Iwasawa, Y. *Journal of the American Chemical Society* **115**(22), 10460–10461 (1993).

[167] Rotzinger, F., Kesselman-Truttmann, J., Hug, S., Shklover, V., and Grätzel, M. *Journal of Physical Chemistry B* **108**(16), 5004–5017 (2004).

[168] Kilså, K. *NCPV and Solar Program Review Meeting 2003* (2003).

[169] Vittadini, A., Selloni, A., Rotzinger, F., and Grätzel, M. *Journal of Physical Chemistry B* **104**(6), 1300–1306 (2000).

[170] Kohle, O., Grätzel, M., Meyer, A. F., and Meyer, T. B. *Advanced Materials* **9**(11), 904–906 (1997).

[171] Wolfbauer, G., Bond, A., Eklund, J., and MacFarlane, D. *Solar Energy Materials and Solar Cells* **70**(1), 85–101 (2001).

[172] Kawano, R. and Watanabe, M. *Chemical Communications* **3**, 330–331 (2003).

[173] Kawano, R., Matsui, H., Matsuyama, C., Sato, A., Susan, M. A. B. H., Tanabe, N., and Watanabe, M. *Journal of Photochemistry and Photobiology A: Chemistry* **164**(1-3), 87–92 (2004).

[174] Wang, P., Zakeeruddin, S., Moser, J., Nazeeruddin, M., Sekiguchi, T., and Grätzel, M. *Nature Materials* **2**(7), 498–498 (2003).

[175] Yoshida, Y., Muroi, K., Otsuka, A., Saito, G., Takahashi, M., and Yoko, T. *Inorganic Chemistry* **43**(4), 1458–1462 (2004).

[176] Fei, Z., Kuang, D., Zhao, D., Klein, C., Ang, W. H., Zakeeruddin, S. M., Grätzel, M., and Dyson, P. J. *Inorganic Chemistry* **45**(26), 10407–10409 (2006).

[177] Berginc, M., Krasovec, U. O., Jankovec, M., and Topic, M. *Solar Energy Materials and Solar Cells* **91**(9), 821–828 (2007).

[178] Yamanaka, N., Kawano, R., Kubo, W., Masaki, N., Kitamura, T., Wada, Y., Watanabe, M., and Yanagida, S. *Journal of Physical Chemistry B* **111**(18), 4763–4769 (2007).

[179] Kato, T. *Science* **295**(5564), 2414–2418 (2002).

[180] Yoshio, M., Kato, T., Mukai, T., Yoshizawa, M., and Ohno, H. *Molecular Crystals and Liquid Crystals* **413**, 2235–2244 (2004).

[181] Born, M. *Zeitschrift für Physik* **1**, 45 (1920).

[182] Riddick, J. and Bunger, W. *Organic Solvents*. Wiley-Interscience, (1970).

[183] Huheey, J., Keiter, E., and Keiter, R. *Anorganische Chemie*. Walter de Gruyter, 2. edition, (1995).

[184] Marcus, Y. *Pure and Applied Chemistry* **59**(9), 1093–1101 (1987).

[185] Weber, R. *Photoelectron Spectroscopy of Liquid Water and Aqueous Solutions in Free Microjets Using Synchrotron Radiation*. PhD thesis, FU Berlin, (2003).

[186] Weber, R., Winter, B., Schmidt, P., Widdra, W., Hertel, I., Dittmar, M., and Faubel, M. *Journal of Physical Chemistry B* **108**(15), 4729–4736 (2004).

[187] Mayer, T. *Photoelektronenspektroskopie an Modellgrenzflächen energiewandelnder Halbleiter/Elektrolyt-Kontakte*. PhD thesis, TU Berlin, (1993).

[188] Markovich, G., Perera, L., Berkowitz, M., and Cheshnovsky, O. *Journal of Chemical Physics* **105**(7), 2675–2685 (1996).

[189] Marcus, R. *Journal of Physical Chemistry* **94**(3), 1050–1055 (1990).

[190] Nozik, A. and Memming, R. *Journal of Physical Chemistry* **100**(31), 13061–13078 (1996).

[191] Zeyer, Grüniger, and Dossenbach. *Journal of Applied Electrochemistry* **22**(3), 304–306 (1992).

[192] Oskam, G., Bergeron, B., Meyer, G., and Searson, P. *Journal of Physical Chemistry B* **105**(29), 6867–6873 (2001).

[193] Nakade, S., Makimoto, Y., Kubo, W., Kitamura, T., Wada, Y., and Yanagida, S. *Journal of Physical Chemistry B* **109**(8), 3488–3493 (2005).

[194] Sapp, S., Elliott, C., Contado, C., Caramori, S., and Bignozzi, C. *Journal of the American Chemical Society* **124**(37), 11215–11222 (2002).

[195] Nusbaumer, H., Zakeeruddin, S. M., Moser, J. E., and Grätzel, M. *Chemistry-A European Journal* **9**(16), 3756–3763 (2003).

[196] Nusbaumer, H., Moser, J., Zakeeruddin, S., Nazeeruddin, M., and Grätzel, M. *Journal of Physical Chemistry B* **105**(43), 10461–10464 (2001).

[197] Clifford, J. N., Palomares, E., Nazeeruddin, M. K., Grätzel, M., and Durrant, J. R. *Journal of Physical Chemistry C* **111**(17), 6561–6567 (2007).

[198] Riedel, E. *Anorganische Chemie*. Walter de Gruyter, 3. edition, (1994).

[199] Calabrese, V. and Khan, A. *Journal of Physical Chemistry A* **104**(6), 1287–1292 (2000).

[200] Liu, Y., Hagfeldt, A., Xiao, X.-R., and Lindquist, S.-E. *Solar Energy Materials and Solar Cells* **55**(3), 267–281 (1998).

[201] Pelet, S., Moser, J., and Grätzel, M. *Journal of Physical Chemistry B* **104**(8), 1791–1795 (2000).

[202] Redmond, G. and Fitzmaurice, D. *Journal of Physical Chemistry* **97**(7), 1426–1430 (1993).

[203] Enright, B., Redmond, G., and Fitzmaurice, D. *Journal of Physical Chemistry* **98**(24), 6195–6200 (1994).

[204] Lyon, L. and Hupp, J. *Journal of Physical Chemistry* **99**(43), 15718–15720 (1995).

[205] Lemon, B. and Hupp, J. *Journal of Physical Chemistry B* **101**(14), 2426–2429 (1997).

[206] Hara, K., Nishikawa, T., Kurashige, M., Kawauchi, H., Kashima, T., Sayama, K., Alka, K., and Arakawa, H. *Solar Energy Materials and Solar Cells* **85**(1), 21–30 (2005).

[207] Agrell, H., Lindgren, J., and Hagfeldt, A. *Solar Energy* **75**(2), 169–180 (2003).

[208] Leon, C., Kador, L., Peng, B., and Thelakkat, M. *Journal of Physical Chemistry B* **109**(12), 5783–5789 (2005).

[209] Stanley, A., Verity, B., and Matthews, D. *Solar Energy Materials and Solar Cells* **52**(1-2), 141–154 (1998).

[210] Angell, C. and Howell, M. *Journal of Physical Chemistry* **73**(8), 2551–2554 (1969).

[211] Knoezinger, H. and Krietenbrink, H. *Journal of the Chemical Society - Faraday Transactions I* **71**, 2421–2430 (1975).

[212] Pelmenschikov, A., Morosi, G., Gamba, A., Coluccia, S., Martra, G., and Paukshtis, E. *Journal of Physical Chemistry* **100**(12), 5011–5016 (1996).

[213] Zhuang, J., Rusu, C. N., and Yates, J. T. *Journal of Physical Chemistry B* **103**(33), 6957–6967 (1999).

[214] Aboulayt, A., Binet, C., and Lavalley, J. *Journal of the Chemical Society - Faraday Transactions* **91**(17), 2913–2920 (1995).

[215] Davit, P., Martra, G., Coluccia, S., Augugliaro, V., Lopez, E., Loddo, V., Marci, G., Palmisano, L., and Schiavello, M. *Journal of Molecular Catalysis A - Chemical* **204**, 693–701 (2003).

[216] Lichtin, N. and Avudaithai, M. *Environmental Science and Technology* **30**(6), 2014–2020 (1996).

[217] Addamo, M., Augugliaro, V., Coluccia, S., Faga, M., Garcia-Lopez, E., Loddo, V., Marci, G., Martra, G., and Palmisano, L. *Journal of Catalysis* **235**(1), 209–220 (2005).

[218] Augugliaro, V., Coluccia, S., Garcia-Lopez, E., Loddo, V., Marci, G., Martra, G., Palmisano, L., and Schiavello, M. *Topics in Catalysis* **35**(3-4), 237–244 (2005).

[219] Rasko, J. and Kiss, J. *Catalysis Letters* **109**(1-2), 71–76 (2006).

[220] Chuang, C., Wu, W., Lee, M., and Lin, J. *Physical Chemistry Chemical Physics* **2**(17), 3877–3882 (2000).

[221] Suda, Y. *Langmuir* **4**(1), 147–152 (1988).

[222] Nagao, M. and Suda, Y. *Langmuir* **5**(1), 42–47 (1989).

[223] Reiß, S., Krumm, H., Niklewski, A., Staemmler, V., and Wöll, C. *Journal of Chemical Physics* **116**(17), 7704–7713 (2002).

[224] Suzuki, S., Yamaguchi, Y., Onishi, H., Sasaki, T., Fukui, K., and Iwasawa, Y. *Journal of the Chemical Society - Faraday Transactions* **94**(1), 161–166 (1998).

[225] Zhou, J., Dag, S., Senanayake, S. D., Hathorn, B. C., Kalinin, S. V., Meunier, V., Mullins, D. R., Overbury, S. H., and Baddorf, A. P. *Physical Review B* **74**(12) (2006).

[226] Lichtin, N. and Sadeghi, M. *Journal of Photochemistry and Photobiology A - Chemistry* **113**(1), 81–88 (1998).

[227] Raza, H., Wincott, P., Thornton, G., Casanova, R., and Rodriguez, A. *Surface Science* **404**(1-3), 710–714 (1998).

[228] Siegbahn, K. *Philosophical Transactions of the Royal Society of London Series A* **268**(1184), 33–57 (1970).

[229] Siegbahn, K. *Pure and Applied Chemistry* **48**(1), 77–97 (1976).

[230] Hertz, H. *Annalen der Physik* **267**, 421–448 (1887).

[231] Hallwachs, W. *Annalen der Physik* **269**, 301–312 (1887).

[232] Einstein, A. *Annalen der Physik* **17**, 132 (1905).

[233] Hüfner, S. *Photoelectron Spectroscopy*. Springer, (1996).
[234] Kibel, M. *X-Ray Photoelectron Spectroscopy*, chapter 7, 165–187. Springer (1992).
[235] Schattke, W., Van Hove, M., Garcia de Abajo, F., Diez Muino, R., and Mannella, N. *Overview of Core and Valence Photoemission*, chapter 2, 50–116. Wiley-VCH (2003).
[236] Hedin, L. *General Theory of Core Electron Photoemission*, chapter 3, 116–141. Wiley-VCH (2003).
[237] Jaegermann, W. *Grenzflächen halbleitender Metallchalkogenide.* professorial dissertation, FU Berlin, (1992).
[238] Rhodin, T. and Gadzuk, J. *The Nature of the Surface Chemical Bond*. North-Holland, Amsterdam, (1979).
[239] Moulder, J., Stickle, W., Sobol, P., and Bomben, K. *Handbook of X-ray Photoelectron Spectroscopy*. Physical Electronics, (1995).
[240] Wagner, C., Naumkin, A., Kraut-Vass, A., Allison, J., Powell, C., and J.R., R. *NIST X-ray Photoelectron Spectroscopy Database;* http://srdata.nist.gov/xps/.
[241] Shirley, D. *Bulletin of the American Physical Society* **18**, 314 (1973).
[242] http://www.specs.de/cms/upload/PDFs/ApplNotes/CasaLetters-Quantification.pdf.
[243] Davis, L. and Feldkamp, L. *Physical Review Letters* **44**(10), 673–676 (1980).
[244] Davis, L. and Feldkamp, L. *Physical Review B* **23**(12), 6239–6253 (1981).
[245] Davis, L. *Physical Review B* **25**(4), 2912–2915 (1982).
[246] Davis, L. *Journal of Applied Physics* **59**(6), R25–R63 (1986).
[247] Kay, A., Arenholz, E., Mun, S., de Abajo, F. J. G., Fadley, C. S., Denecke, R., Hussain, Z., and Van Hove, M. A. *Science* **281**(5377), 679–683 (1998).
[248] Nerlov, J., Ge, Q. F., and Moller, P. J. *Surface Science* **348**(1-2), 28–38 (1996).
[249] Tezuka, Y., Shin, S., Agui, A., Fujisawa, M., and Ishii, T. *Journal of the Physical Society of Japan* **65**(1), 312–317 (1996).
[250] Raman, C. V. and Krishnan, K. *Nature* **121**(3048), 501–502 (1928).
[251] Bragg, W. *Nature* **90**, 410 (1912).
[252] Binnig, G., Quate, C., and Gerber, C. *Physical Review Letters* **56**(9), 930–933 (1986).
[253] http://de.wikipedia.org/wiki/Rasterkraftmikroskop.
[254] http://www.tu-darmstadt.de/fb/ms/fg/ofl/methoden/AFM/afm_theorie.htm.
[255] Ashcroft, N. and Mermin, N. *Festkörperphysik*. Oldenbourgh Verlag, (2001).
[256] Chen, C. *Introduction to Scanning Tunneling Microscopy*. Oxford University Press, (1993).
[257] von Ardenne, M. *Zeitschrift für Physik A - Hadrons and Nuclei* **109**(9), 553–572 (1938).

[258] http://www.solaronix.com/technology/assembly.
[259] Sandell, A., Anderson, M. P., Alfredsson, Y., Johansson, M. K. J., Schnadt, J., Rensmo, H., Siegbahn, H., and Uvdal, P. *Journal of Applied Physics* **92**(6), 3381–3387 (2002).
[260] Sandell, A., Andersson, M. P., Johansson, M. K. J., Karlsson, P. G., Alfredsson, Y., Schnadt, J., Siegbahn, H., and Uvdal, P. *Surface Science* **530**(1-2), 63–70 (2003).
[261] Fictorie, C. P., Evans, J. F., and Gladfelter, W. L. *Journal of Vacuum Science and Technology A - Vacuum Surfaces and Films* **12**(4), 1108–1113 (1994).
[262] Mayer, T., Lebedev, M., Hunger, R., and Jaegermann, W. *Applied Surface Science* **252**(1), 31–42 (2005).
[263] Niilisk, A., Moppel, M., Pars, M., Sildos, I., Jantson, T., Avarmaa, T., Jaaniso, R., and Aarik, J. *Central European Journal of Physics* **4**(1), 105–116 (2006).
[264] Shultz, A., Jang, W., Hetherington, W., Baer, D., Wang, L., and Engelhard, M. *Surface Science* **339**(1-2), 114–124 (1995).
[265] Schwanitz, K., Weiler, U., Hunger, R., Mayer, T., and Jaegermann, W. *Journal of Physical Chemistry C* **111**(2), 849–854 (2007).
[266] Mayer, J. T., Diebold, U., Madey, T. E., and Garfunkel, E. *Journal of Electron Spectroscopy and Related Phenomena* **73**(1), 1–11 (1995).
[267] Gouttebaron, R., Cornelissen, D., Snyders, R., Dauchot, J. P., Wautelet, M., and Hecq, M. *Surface and Interface Analysis* **30**(1), 527–530 (2000).
[268] Mönch, W. *Semiconductor Surfaces and Interfaces*. Springer New York, 3 edition, (1995).
[269] Powell, C. *NIST Electron Inelastic-Mean-Free-Path Database*. computer program.
[270] Xin, B., Jing, L., Ren, Z., Wang, B., and Fu, H. *Journal of Physical Chemistry B* **109**(7), 2805–2809 (2005).
[271] Bullock, E. L., Patthey, L., and Steinemann, S. G. *Surface Science* **352-354**, 504–510 (1996).
[272] Sham, T. and Lazarus, M. *Chemical Physics Letters* **68**(2-3), 426–432 (1979).
[273] Kitao, O. *Journal of Physical Chemistry C* **111**(43), 15889–15902 (2007).
[274] Yeh, J. and Lindau, I. *Atomic Data and Nuclear Data Tables* **32**(1), 1–155 (1985).
[275] Berger, T., Sterrer, M., Diwald, O., Knozinger, E., Panayotov, D., Thompson, T. L., and Yates, J. T. *Journal of Physical Chemistry B* **109**(13), 6061–6068 (2005).
[276] Berger, T., Sterrer, M., Stankic, S., Bernardi, J., Diwald, O., and Knozinger, E. *Materials Science and Engineering C* **25**(5-8), 664–668 (2005).
[277] Kurtz, R. L. and Henrich, V. E. *Physical Review B* **25**(6), 3563–3563 (1982).
[278] Tanuma, S., Powell, C., and Penn, D. *Surface amd Interface Analysis* **21**(3), 165–176 (1994).
[279] Holland, D. and Karlsson, L. *Journal of Electron Spectroscopy and Related Phenomena* **150**(1), 47–55 (2006).

[280] Borodin, A., Hofft, O., Kahnert, U., Kempter, V., and Allouche, A. *Vacuum* **73**(1), 15–24 (2004).

[281] Bieri, G., Heilbronner, E., Hornung, V., Klosterjensen, E., Maier, J., Thommen, F., and Niessen, W. *Chemical Physics* **36**(1), 1–14 (1979).

[282] Borodin, A., Hofft, O., Kahnert, U., Kempter, V., Krischok, S., and Abou-Helal, M. O. *Journal of Chemical Physics* **120**(11), 5407–5413 (2004).

[283] Rensmo, H., Westermark, K., Södergren, S., Kohle, O., Persson, P., Lunell, S., and Siegbahn, H. *Journal of Chemical Physics* **111**(6), 2744–2750 (1999).

[284] Thiel, P. and Madey, T. *Surface Science Reports* **7**(6-8), 211–385 (1987).

[285] Feibelman, P., Hamann, D., and Himpsel, F. *Physical Review B* **22**(4), 1734–1739 (1980).

[286] Stockbauer, R., Hanson, D., Flodstrom, S., and Madey, T. *Journal of Vacuum Science and Technology* **20**(3), 562–563 (1982).

[287] Stockbauer, R., Hanson, D., Flodstrom, S., and Madey, T. *Physical Review B* **26**(4), 1885–1892 (1982).

[288] Mayer, T. and Jaegermann, W. *Journal of Physical Chemistry B* **104**(25), 5945–5952 (2000).

[289] Brookes, I. M., Muryn, C. A., and Thornton, G. *Physical Review Letters* **87**(26) (2001).

[290] Kurtz, R., Stockbauer, R., Madey, T., Roman, E., and Desegovia, J. *Surface Science* **218**, 178–200 (1989).

[291] DiValentin, C., Tilocca, A., Selloni, A., Beck, T., Klust, A., Batzill, M., Losovyj, Y., and Diebold, U. *Journal of the American Chemical Society* **127**, 9895–9903 (2005).

[292] Liu, G., Klein, A., Thissen, A., and Jaegermann, W. *Surface Science* **539**, 37–48 (2003).

[293] Rensmo, H., Sodergren, S., Patthey, L., Westermark, K., Vayssieres, L., Kohle, O., Bruhwiler, P. A., Hagfeldt, A., and Siegbahn, H. *Chemical Physics Letters* **274**(1-3), 51–57 (1997).

[294] Karlsson, P. G., Bolik, S., Richter, J. H., Mahrov, B., Johansson, E. M. J., Blomquist, J., Uvdal, P., Rensmo, H., Siegbahn, H., and Sandell, A. *Journal of Chemistry Physics* **120**(23), 11224–32 (2004).

[295] Schwanitz, K., Mankel, E., Hunger, R., Mayer, T., and Jaegermann, W. Technical report, Berliner Elektronenspeicherring-Gesellschaft für Synchrotronstrahlung (BESSY), Berlin, (2006).

[296] Agrell, H. G., Lindgren, J., and Hagfeldt, A. *Journal of Photochemistry and Photobiology A - Chemistry* **164**(1-3), 23–27 (2004).

[297] Greijer, H., Lindgren, J., and Hagfeldt, A. *Journal of Physical Chemistry B* **105**(27), 6314–6320 (2001).

[298] Nahon, L., Duffy, L., Morin, P., Combetfarnoux, F., Tremblay, J., and Larzilliere, M. *Physical Review A* **41**(9), 4879–4888 (1990).

[299] Nahon, L., Svensson, A., and Morin, P. *Physical Review A* **43**(5), 2328–2337 (1991).

[300] Amusia, M., Cherepkov, N., Chernysheva, L., and Manson, S. *Physical Review A* **6102**(2) (2000).

[301] Bournel, F., Gallet, J., Kubsky, S., Dufour, G., Rochet, F., Simeoni, M., and Sirotti, F. *Surface Science* **513**(1), 37–48 (2002).

[302] Rangan, S., Bournel, F., Gallet, J. J., Kubsky, S., Le Guen, K., Dufour, G., Rochet, F., Sirotti, F., Carniato, S., and Ilakovac, V. *Physical Review B* **71**(16), 165319/1–165319/11 (2005).

[303] Kishi, K., Okino, Y., and Fujimoto, Y. *Surface Science* **176**(1-2), 23–31 (1986).

[304] Markovich, G., Pollack, S., Giniger, R., and Chesnowsky, O. *Journal of Chemical Physics* **101**(11), 9344–9353 (1994).

[305] Boschloo, G., Haggman, L., and Hagfeldt, A. *Journal of Physical Chemistry B* **110**(26), 13144–13150 (2006).

[306] Dürr, M., Yasuda, A., and Nelles, G. *Applied Physics Letters* **89**(6), 061110/1–061110/3 (2006).

Herzlichen Dank an...

- **Prof. Dr. Wolfram Jaegermann** für die freundliche Aufnahme in sein Arbeitsgebiet und für die positive Arbeitsatmosphäre in seiner Arbeitsgruppe.
- **Prof. Dr. Wolgang Ensinger** für die Übernahme des Zweitgutachtens.
- **Dr. Thomas Mayer** für die hervorragende Betreuung während der gesamten Doktorarbeitszeit. Weiterhin schätzte ich seine Kreativität bei der Entwicklung neuer Ideen, seiner unermüdlichen Bereitschaft zur Diskussion und das Vermitteln einer positiven (!) Streitkultur.
- **Eric Mankel** für seine äußerst hilfreiche und tapfere Unterstützung bei den BESSY Messzeiten und sein Improvisationstalent, insbesonders bei terminlichen Engpässen bezüglich Zimmerbelegung beim BESSY-Nutzertreffen.
- **Dr. Ralf Hunger** und **Wolfgang Bremsteller** für die tatkräftige Hilfe auch zu ungünstigsten Tages- und Nachtzeiten bei allen technischen Problemen, welche mit der SoLiAS bei BESSY zusammenhingen.
- **Dr. Patrick Hoffmann** für die Instandhaltung der U49/PGM-2 Beamline bei BESSY
- **Dr. Christian Pettenkofer** für die Instandhaltung der TGM7 Beamline bei BESSY
- **Dr. Ulrich Weiler** für die fruchtbare Zusammenarbeit während der ersten beiden BESSY-Messzeiten.
- **Gabi Haindl** für die Instandhaltung des integrierten UHV Systems DaISy-Sol.
- **Marga Lang** für die Erledigung einer Vielzahl bürokratischer Dinge.
- **Johannes Luschitz** für die REM-Aufnahmen.
- **Thorsten Enz** und **Sebastian Gottschalk** von der Arbeitsgruppe Gemeinschaftslabor Nanomaterialien für die GIXRD Messungen.
- **Dr. David Ensling**, **Dr. Frank Säuberlich** und **Dr. Bengt Jäckel** für die Beantwortung dringender Fragen während des Zusammenschreibens via Skype.
- **Johannes Türck** für die AFM-Messungen.
- **Dr. Jörg Rappich** vom Hahn Meitner Institut für die Benutzung des Raman Spektroskops.
- **die Mitarbeiter der Werkstatt** für die präzise Anfertigung auch kurzfristig benötigter Teile.
- **meine Frau Yun-Young** für ihre unendliche Geduld, besonders während der Zeit des Zusammenschreibens.
- 저를 항상 격려 해주신 한국에 계신 가족 여러분께 감사 드립니다.
- **meine Eltern**, welche mich während meines gesamten Studiums unterstützten und daher diese Arbeit ermöglichten.
- alle namentlich nicht genannten, welche mir in irgendeiner Weise während dieser Zeit weitergeholfen haben.

Die VDM Verlagsservicegesellschaft sucht für wissenschaftliche Verlage abgeschlossene und herausragende

Dissertationen, Habilitationen, Diplomarbeiten, Master Theses, Magisterarbeiten usw.

für die kostenlose Publikation als Fachbuch.

Sie verfügen über eine Arbeit, die hohen inhaltlichen und formalen Ansprüchen genügt, und haben Interesse an einer honorarvergüteten Publikation?

Dann senden Sie bitte erste Informationen über sich und Ihre Arbeit per Email an *info@vdm-vsg.de*.

Sie erhalten kurzfristig unser Feedback!

VDM Verlagsservicegesellschaft mbH
Dudweiler Landstr. 99
D - 66123 Saarbrücken

Telefon +49 681 3720 174
Fax +49 681 3720 1749

www.vdm-vsg.de

Die VDM Verlagsservicegesellschaft mbH vertritt

Printed by Books on Demand GmbH, Norderstedt / Germany